Concrete Mix Design, Quality Control and Specification

Fourth Edition

Concrete Mix Design, Quality Control and Specification

Fourth Edition

Ken W. Day
James Aldred
Barry Hudson

CRC Press
Taylor & Francis Group
Boca Raton   London   New York

CRC Press is an imprint of the
Taylor & Francis Group, an **informa** business

A SPON PRESS BOOK

CRC Press
Taylor & Francis Group
6000 Broken Sound Parkway NW, Suite 300
Boca Raton, FL 33487-2742

First issued in paperback 2017

Version Date: 20130918

ISBN 13: 978-0-415-50499-7 (hbk)
ISBN 13: 978-1-138-07353-1 (pbk)

Library of Congress Cataloging-in-Publication Data

Day, Ken W.
    Concrete mix design, quality control and specification / Ken W. Day, James Aldred, Barry Hudson. -- Fourth edition.
        pages cm
    Includes bibliographical references and index.
    ISBN 978-0-415-50499-7 (hardback)
    1. Concrete--Mixing--Quality control. 2. Concrete--Specifications. I. Aldred, James. II. Hudson, Barry. III. Title.

TA439.D39 2014
620.1'36--dc23   2013032492

Visit the Taylor & Francis Web site at
http://www.taylorandfrancis.com

and the CRC Press Web site at
http://www.crcpress.com

Contents

Acknowledgements

Acknowledgements for this fourth edition are largely covered in the introduction but, as with the third edition, we do not wish earlier influences to be forgotten and so we repeat the acknowledgements of the second and third editions.

ACKNOWLEDGEMENTS TO THE THIRD EDITION

My third edition builds upon the shoulders of the work done for the first two, and I do not wish those I thanked then to be forgotten now. Therefore the acknowledgements in the second edition are reprinted in full following those for the current edition.

My company, Concrete Advice Pty Ltd, was sold in 2001 to Maricopa Readymix, my first U.S. client, at the instigation of Dave Hudder, at that time managing director of Maricopa. I have him to thank for his recognition of the value of ConAd in the United States and for providing me with the means to enjoy my semiretirement and to travel the world preaching my concepts.

Upon Dave leaving Maricopa, Concrete Advice was on-sold to Command Alkon. I was very pleased about this because ConAd is a perfect fit for a major, worldwide batching-system provider. I thank them for continuing my part-time consultancy until the end of 2004, even though I have had little influence on the new version of ConAd.

I will never forget the part played by Don Bain, technical manager of Maricopa, in all of this. It was he who recommended the initial purchase of the ConAd system to Dave Hudder back in early 2000, he who used ConAd to enable the expansion of Maricopa and build the U.S. reputation of ConAd, and he who left Maricopa for a time to help Command Alkon with initial marketing of ConAd. His written contribution to this text is appreciated, but it is negligible compared to his contribution to the reputation of the ConAd system.

Andrew Travers continues to labor prodigiously as CEO of ConAd. Its future now depends on him as he rushes around the world promoting and installing it. Unfortunately he has been far too busy to write a section of this book, much as he wanted to, and much as it would have been appreciated. Perhaps he will write the next edition.

Two other stalwarts, having contributed greatly, are no longer able to do so. Dan Leacy, the Australian equivalent of Don Bain, unfortunately passed away at an early age, and Michael Shallard retired at an even earlier age after a severe illness, depriving the system of its major source of computer expertise. I shall remember them.

My e-mail directory overflows with large numbers of people substituting for my lack of field experience in recent years. Several names appear as contributing sections of the text: Dr Alex Leshchinsky and his father Dr Marat Lesinskij, Mark Mackenzie, Dr Norwood Harrison, Dr Grant Lukey, John Harrison, Tracy Goldsworthy, and Dr Joe Dewar, whose contribution to the previous edition is repeated here.

Contributions not so acknowledged, but nevertheless real, include Aulis Kappi, Charles Allen, James Aldred, Kevin Galvin, Lawrence Roberts, Richard Hall, Dr Celik Ozylidirim, Jay Lukkarila, Dr Steve Trost, also Barry and Tania Hudson for their magnificent forum on the website http://www.aggregateresearch.com. It should be emphasised that several of these do not agree with all that I have written, so any credit for the work is shared with them, but any blame is mine alone.

Justin Smyth (delphian@smythconsulting.net) operates my website (http://www.kenday.id.au) and has amended the free programs on that site.

ACKNOWLEDGEMENTS TO THE SECOND EDITION

There are three individuals without whom this book could not have happened and four more without whom it may have been very different. The first group comprises: O. Jan Masterman, technical director, Unit Construction Co., London in the 1950s, who somehow inspired and guided me to originate in my first two years of employment the greater part of the philosophy and concepts herein recorded; John J. Peyton, John Connell & Associates (now Connell Wagner), Melbourne, without whose encouragement I never would have started my company Concrete Advice Pty Ltd in 1973 and so the nascent control techniques never would have developed to fruition; John Wallis, formerly Singapore director of Raymond International (of Houston, Texas), without whom my Singapore venture would have foundered in 1980, leaving me without computerisation and without the broad international proving grounds for the mix design system.

The second group comprises John Fowler, who wrote the first computer program using my mix design methods, at a time when I had a firm opinion that mix design was partly an art and could never be computerised; D. A. Stewart, whose book *The Design and Placing of High Quality*

Concrete (Spon, 1951) was a first major influence; David C. Teychenne, who led where I have followed in specific surface mix design; and my son Peter, who transformed "ConAd" from an amateur spreadsheet into a professional computer program.

A third kind of indebtedness is to those who assisted in the actual production of the book. They have become too numerous to list all of them by name but Hasan Ay and Andrew Travers are especially thanked for their work on figures and tables.

Harold Vivian, Bryant Mather, Dr Alex Leshchinsky, and Dr Francois de Larrard are especially thanked for invaluable advice and contributions; Sandor Popovics for his published works and thought-provoking discussions; Joe Dewar, Bryant Mather, and John Peyton for their kind forewords; also Vincent Wallis on whom I have relied for an (often brutally) honest opinion over more than 30 years; and of course my wife, who has endured a great deal in the cause of concrete technology.

A new kind of indebtedness is to those individuals in my major client companies who have not only enabled my company (Concrete Advice Pty Ltd) to survive and prosper but have also contributed in no small measure to improvements in the system. They include Peter Denham and Dan Leacy of CSR Readymix, Paul Moses of Boral, and Mark Mackenzie of Alpha, South Africa.

The ConAd computer program has come a long way since the first edition and thanks are due to my staff at Concrete Advice. Michael Shallard and Lloyd Smiley wrote the latest program and Andrew Travers, now manager of the company, knows how to use it better than I.

Finally I must thank my younger son, John Day, now technical manager of Pioneer Malaysia, for using these techniques so effectively as to make the world's tallest building, Petronas Towers, the best example yet of low variability, high strength concrete.

Introduction

The rapidly changing scene in concrete technology necessitates this fourth edition. Obviously I am aware of these changes, but being retired from active participation in concrete production and control, I have brought in two carefully selected coauthors in addition to obtaining input from many people I consider to be leading experts in their fields.

The most fundamental change is the recognition that water to cement (w/c) ratio is not the best available criterion of quality and durability. This, combined with greenhouse gas and sustainability considerations, has caused cement replacement materials to be viewed in a new light. In the future, little, if any, concrete will be produced without at least one component of this large range of materials.

Diminishing availability of natural sand conforming to preconceived ideal gradings has opened up the field for crusher fines, creating a new imperative to better understand their production and use.

It is not surprising that higher strengths and higher heights of pumpability are available, or that self-compacting concrete is becoming popular—and there will always be new admixtures.

One consequence of the greatly expanded range of materials is that theoretical mix design has essentially become only a tool for the education of new entrants to the field. Practical mixes in use will be the result of feedback, adjustment, and trial and error—but the processes used to accomplish this will be very organised and precise rather than ad hoc.

Ideal quality control (MMCQC), on the other hand, will not change from the ideal described in previous editions. The difference here is that the principles (and practice) set out many times over the years are at last showing signs of universal acceptance, even in the United States. Concrete production must be controlled by the producer.

The specification of concrete will become detailed and precise for other than very routine use. However, it will be detailed and precise in terms of required properties and performance rather than constituents. A specification for a major project is likely to be negotiated and agreed upon rather than imposed. We can look forward to a time when every significant producer will have a range of mixes with well-established properties from which a selection can be made.

Durability is a major topic; having discarded w/c ratio as the best criterion, a new criterion must be found. This needs to be in the form of a physical test because it must be applicable to a wide range of different formulations. Although we are concerned with durability for an extended life, a test at as early an age as possible is needed to form part of the QC process. Although strength can no longer be regarded as a criterion of durability, it retains its importance as a detector of change. A change point in strength is a change point in the mix quality and so may be the earliest way of detecting change in durability. However, having detected a change, its effect on durability cannot be established on strength grounds and a specific durability test is needed.

Ken W. Day
Nunawading, Australia

About the Authors

Ken W. Day is well known as the author of the first three editions of this book. He has worked continuously in concrete mix design and quality control since graduating in 1952, except for a short period in the 1960s as associate partner of Harris & Sutherland in the United Kingdom where he worked on battery precasting of concrete housing. His two most important developments have been multigrade, multivariable, cusum quality control and specific surface mix design. He worked initially in the United Kingdom, then Australia, before starting his own company, Concrete Advice Pty Ltd, in 1973. Day has lectured in 23 countries, leading to international use of his concepts, and received multiple international awards for his work, details of which can be read on his website, http://www.kenday.id.au. His company was sold to Command Alkon via Maricopa RMC in 2002 and his quality control (QC) program is now marketed as CommandQC, and is no longer under his control. However, a small program, "KensQC", is available on his website and enables new users to experience his techniques on their own data.

James Aldred has over 30 years of experience in the concrete industry in Australia, Asia, the Middle East, and the United Kingdom. His background includes technical director of an international admixtures company, manager of the High Performance Concrete Research Group at the National University of Singapore, technical manager of Taywood Engineering, and honorary research fellow at Imperial College. He was the independent verifier for the Burj Khalifa in Dubai, which is the world's tallest tower. Dr Aldred is currently technical director with AECOM and an adjunct associate professor at the University of New South Wales. Dr Aldred obtained his PhD from Curtin University (Australia). He is a fellow of the Institute of Engineers Australia, the American Concrete Institute, and the Institute of Concrete Technology, as well as being a LEED Accredited Professional. Dr Aldred has received the Award of Excellence from the Concrete Institute of Australia, an award for outstanding and sustained contributions to concrete technology by ACI International Conferences, and the prestigious George Stephenson Medal from Institute of Civil Engineers.

Barry Hudson has 30 years of experience in the construction materials industry. He currently holds dual roles in Heidelberg Cement, establishing and managing the Competence Center Materials for the TEAM Region (Africa, Northern Europe, Baltics, Benelux, the United Kingdom, and the Mediterranean Basin), an operational overview and best practice organisation that oversees 70 million tonnes of aggregates production and 17 million cubic metres of concrete. Along with this role, he is also director of aggregates for Norway, Sweden, and the Baltics.

Specialising in aggregates for concrete, Hudson has sat on various standards committees around the world as well as industry association representation. During his 11 years of experience in the United States, Hudson was involved with the International Center for Aggregates Research. He was also an integral part of Lafarge research efforts in concrete and aggregates.

Founder of Aggregate Research (http://www.aggregateresearch.com), Hudson has four patents based around concrete mix design and aggregates characterisations. He has been published 37 times and has given presentations on every continent. Hudson is recognised as a person who combines leading edge science with the serious practicalities of everyday production.

Chapter 1

Advice to specifiers

The old adage that no one is more difficult to teach than those who are convinced they already know it all is nowhere more apparent than in the specification and control of concrete quality. It explains why two of the world's most respected sources of knowledge about concrete—American Concrete Institute (ACI) and United Kingdom/Europe—are lagging almost 30 years behind developments in Australia.[1]

Structural designers in concrete are certainly expected and entitled to specify the properties they have assumed in their design, including such items as strength, shrinkage, and resistance to anticipated sources of deterioration. Unfortunately few structural designers are also expert concrete technologists and may be reluctant to admit this. There is a tendency to assume that if you do not have much detailed knowledge of a subject, then there is not much to know about it.

1.1 MIX SELECTION

Existing codes accept that concrete strength tends to be a normally distributed variable and therefore needs to be considered in terms of mean strength and standard deviation rather than an absolute limit. Instances of lower than specified strength should be dealt with by analysing a number of recent results to determine whether the low result constitutes a genuine downturn or an isolated statistical aberration. An investigation by coring or otherwise may be undertaken to confirm the diagnosis.

Although early age strengths may be determined, correction normally tends to be based on a number of 28day results. In the case of the United Kingdom a technique known as a V-mask may be applied to a cusum graph of strength. A CUSUM graph (see Chapter 10) is a more sensitive detector of change and a V-mask automatically applies a precise statistical test to detect a significant downturn. There are actually two basic requirements

---

[1] Have yet to adopt performance-based measures adopted in Australia almost 30 years ago.

of a control system. One is that it should provide a precise judgement of quality and the other that it should react as quickly as possible to restore the required quality in the event of any downturn. Any attempt to combine these two into a single requirement is almost certain to achieve neither.

What needs to be learned is that mix design and quality control must be in the hands of the concrete producer. This has always been the case because any external supervisor cannot require corrective action based on as little evidence as a properly motivated producer will require. It is now even more the case because a large variety of admixtures and cement replacement supplementary cementitious materials is available, and a producer needs to have conducted trials to establish which materials and which suppliers of materials will best enable him or her to consistently produce the most economical satisfactory compliant concrete for the project at hand. It is important that feedback and cooperation be established between the concrete producer and his or her material suppliers, and it is generally undesirable that such links be unnecessarily discarded, along with current performance history, by requiring the use of unfamiliar materials. However, those few specifiers who do have expert knowledge beyond that of most producers should certainly make it available as advice but preferably with an alternative performance option.

In the United States the "P2P" battle still rages (P2P, prescription to performance as a specification basis) and has not yet influenced official ACI practice, even though it is now 10 years since Command Alkon purchased Ken Day's ConAd system and licensed it to hundreds of producers in the United States and around the world.

In the United Kingdom and Europe the current system is nearer to Australian practice but is even more solidly entrenched. The problems there, as presented in Day's paper at the International Federation for Structural Concrete (FIB) 2008, are a limited ability to include multiple groups in a multigrade analysis, a failure to use multivariable CUSUMs to link cause and effect, and postponing problem detection to an analysis of a substantial number of 28-day results.

A genuine difficulty with performance specifications is the difficulty in specifying durability. Opponents of performance specifications see this as a justification for prescribing some mix features.

1.2 SPECIFYING DURABILITY

The common practice to specify a minimum cementitious content to achieve "durability" is misguided. First, Buenfeld and Okundi (1998) showed that, at a given water to cement ratio (w/c), the higher binder content actually increased chloride ion ingress in concrete. This is hardly surprising when transport processes occur primarily through the paste fraction of the concrete. Second, an unnecessarily high cementitious content may lead to increased cracking due to thermal stresses and shrinkage, which could reduce durability. Unnecessarily wasting cementitious materials also increases the environmental impact of the concrete.

A difficulty exists in specifying durability performance owing to the absence of a generally accepted comprehensive test at a reasonably early age. It also depends on the nature of the deteriorating influences to be withstood.

An increasing number of specifications require compliance testing of concrete transport properties during construction in an attempt to improve the expected durability of reinforced concrete structures. However, unlike compressive strength, there is little information available on the expected variation in the results obtained as well as on the relationship between such compliance tests and in situ properties/performance. Indeed, unlike air entrainment to enhance freeze–thaw resistance, the required performance for the different specified parameters to achieve the desired durability has often not been established.

In the case of chloride-induced corrosion, which is the most common durability issue, performance requirements may include diffusion, migration, resistivity or water transport measurements, or combinations of these. In the United States, the ASTM C1202 coulomb test, has been common. The leading contenders for adoption are the coulomb test, chloride migration, and the direct measurement of moisture absorption.

The coulomb test, often called the rapid chloride permeability test even though it is does not measure chlorides or permeability, is a measurement of saturated resistivity and is correlated to chloride diffusion. The test result has quite high variability, so that it should not be specified as a rejection criterion for the sampled concrete. However, a statistical analysis of results over the course of a month could be used as a basis for a penalty clause. An important breakthrough in specification practice is an article on "End Result Specifications" by the Virginia Department of Transportation (DOT) in Concrete International for March 2011. This use of performance specifications, complete with bonus and penalty clauses, by a major U.S. government department could lead to a rapid transformation in world practice.

Chloride diffusion is perhaps the most relevant test, but it is expensive and time consuming to test and therefore not well suited for compliance testing. Chloride migration is a much faster and cheaper procedure that still measures chloride penetration. The recently released RMS B80 specification uses both chloride diffusion and migration values for different chloride environments. We suggest that the best procedure would be to measure resistivity frequently and chloride migration occasionally to confirm adequate performance based on the concept of a characteristic value.

Where chloride ingress is controlled by limting absorption, absorption tests are appropriate, i.e., BS1881: P_{t122}.

1.3 SPECIFYING THERMAL LIMITS

Specifications for temperature rise and differentials in massive pours require attention. A default peak temperature of 70°C is prudent as it would virtually eliminate the possible problem of delayed ettringite formation. Many specifiers focus on the temperature differential within the concrete mass and a value of 20°C is often specified. Most concrete with crushed aggregate can tolerate a much higher differential without cracking, 28°C would probably be a better default value. However, in our experience, most thermal cracking has been caused by external restraint of massive concrete elements by a rigid substrate during cooling. The attention on the differential temperature requirement in temperate conditions often leads to excessive insulation and increases both the peak temperature and the volume of concrete that reached high temperature. Therefore, to reduce a minor potential problem, the more likely problem is exacerbated.

1.4 SPECIFYING RHEOLOGY

There is a tendency to limit concrete workability in specifications based on the assumption that lower workability produces better concrete. Although often true when added water was the only way to increase workability, it is certainly not true in the age of advanced admixtures. Poor workability can lead to honeycombing, slower construction, and uncontrolled water addition after compliance sampling. Resultant defects can lead to costly repairs and litigation where the specification will come under scrutiny. The problem of prescriptive specification of rheology can also occur with self-consolidating concrete (SCC) where overzealous specifiers can require very high workability parameters, which can lead to segregation. We would suggest that the specification require that the contractor/premix company confirm that the rheology of the concrete is satisfactory for the proposed placement procedure and the mix developed complies with the performance parameters. Site compliance testing would be used to confirm that the supplier complied with their agreed rheology.

1.5 SPECIFYING DRYING SHRINKAGE

Many specifications include limits on drying shrinkage according to a standard procedure such as ASTM C157 or AS 1012.13. Although this may seem prudent and would be expected to reduce cracking, there are a number of dangers of specifying stringent drying shrinkage limits. The test procedures are conducted on well-cured small specimens 75 mm × 75 mm (3 inches × 3 inches) in cross-section dried at 50% relative humidity and therefore not representative of drying of standard concrete elements exposed to drying in most environments. Higher strength concrete with higher cementitious contents tends to exhibit lower shrinkage in these tests. However, such mixes may have greater movement due to higher peak temperatures and more autogenous shrinkage, which are not measured in the test.

We would recommend using models such as CIRIA C660 to help determine if shrinkage is a problem and specify a shrinkage limit or a shrinkage-reducing admixture when necessary. Curing by water ponding or the use of saturated lightweight aggregate are good ways to limit or even eliminate autogenous shrinkage.

Chapter 2

Cementitious materials

2.1 PORTLAND CEMENT

2.1.1 Introduction

No attempt is made in this book to provide a general background and description of Portland cement. Such information is available in almost any textbook on concrete as well as many specialised books on cement. A particularly recommended reference is the American Concrete Institute (ACI) Guide to the Selection and Use of Hydraulic Cements (ACI 225R, 1999). This is a very comprehensive 29-page dissertation with an equally comprehensive list of further references. Another useful reference is *High Performance Concrete* (Aitcin, 2011), which provides substantial detail on cement, and also on cementitious materials and admixtures.

What is attempted in the current section is a guide to the extent to which changes in concrete properties may be due to changes in the cementitious material used.

What can go wrong with cement?

1. As the user experiences it
   a. Setting—It can set too quickly or too slowly.
   b. Strength development—It can develop less strength than usual.
   c. Water requirement and workability—It can have a higher water requirement or act as a less suitable lubricant than usual.
   d. Bleeding—It can inhibit bleeding less successfully or at the other extreme produce a "stickier" mix than usual.
   e. Disruptive expansion.
   f. Reduced chemical resistance.
   g. Too rapid evolution of heat.
   h. Deterioration in storage (either before of after grinding).
   i. It can arrive hot, that is, hotter than usual increasing concrete placement temperature.

    j. It can be delivered from the same depot and even ground at the same plant but be produced from a different clinker, that is, imported clinker using different materials and produced in a different kiln may have been used.

    k. Sometimes the mill certificate with the cement does not relate to the actual cement delivered.

2. As it is produced:

    a. Variation in raw materials.

    b. Segregation at any of several stages.

    c. Incorrect proportion or uneven distribution of gypsum ($CaSO_4.2H_2O$).

    d. Variable firing and grinding temperatures.

    e. Unsatisfactory grinding, including overall fineness, particle size distribution, and particle shape.

    f. Deterioration (including segregation) of clinker in storage.

    g. Seasonal variations.

2.1.2 Significant test results

Cement users in some parts of the world can obtain test certificates from their cement suppliers. The following may be of assistance in interpreting the kind of information usually provided on such certificates. Where no test data are obtained in this way, it may be considered too expensive to undertake routine testing on behalf of a single project or small ready-mix plant. A solution to this problem is to take a sample either daily or from each truck of cement (whichever is least). The sample should be kept in a (well-labeled!) sealed container until the 28-day concrete test results are obtained and then discarded. A sample is then available and should be tested if unsatisfactory concrete test results are encountered for which no other explanation can be found.

    Where regular test data are obtained, it is useful to maintain graphs of the information provided. As with concrete test data, cusum (cumulative sum) graphs are far more effective at detecting change points. (See Chapter 10.)

    The main results likely to be provided are

1. Setting time—Initial and final set are both arbitrary stages in smooth curve of strength development. Abnormal results can indicate incorrect proportion of gypsum, excessive temperature in final grinding (which dehydrates gypsum and alters its effectiveness) or deterioration with age.

2. Fineness, finer cement will

    a. React more quickly (faster heat generation)

    b. React more completely

   c. Improve mix cohesion (or make "sticky")

   d. Reduce bleeding

   e. Deteriorate more quickly

   f. Be more susceptible to cracking

   g. Generally require more water (note that this may be less due to any direct effect of fineness than to the reduced range of particle sizes normally resulting from finer grinding)

3. Soundness (Pat, Le Chatelier, and autoclave tests) is intended to detect excessive free lime (perhaps due to incomplete blending rather than wrong chemical proportions). Some experts disagree that the intention is achieved, but this is beyond the present scope. Magnesia can also cause unsoundness (if as periclase) but perhaps too slowly for Pat or Le Chatelier, needs autoclave or chemical limit (and see Section 13.2.6.5 for intentional use of a proportion of magnesia).

4. Normal consistency—Generally just a starting point for other tests but can show up undesirable grinding characteristics. Where very high strength concrete is involved, large amounts of cement will be required and a very low w/c ratio will be required. A cement with a high water demand is a disadvantage in such circumstances. Interesting uses for this test are as a compatibility check between admixtures and cement or to determine the effect on water requirement of a percentage of fly ash or silica fume.

5. Loss on ignition—Mainly a check on deterioration during storage. The test drives off any moisture or carbon dioxide that may have been absorbed. A 3% loss on ignition could mean a 20% strength loss if this was due to moisture. However, up to 5% of limestone ($CaCO_3$) is permitted to be added to cement and this test would drive off CO_2 from limestone.

6. Sulfur trioxide/sulfide (SO_3) check on proportion of gypsum has considerable significance for setting time, strength development and shrinkage. The test determines the content of SO_3 from all sources (e.g., added gypsum, oxidised sulfur in fuels, etc.) and in all states. It therefore may not be an accurate guide to the amount of active (soluble) SO_3 present. It is the amount of active SO_3, which affects, for example, setting time, rate of strength development, and tendency for shrinkage and cracking.

7. Insoluble residue—Check on impurities or nonreactive content only, the effect is the same as reducing the cement content by the percentage of the insoluble material. However, this test may characterise fly ash as insoluble residue and any limit should be based on the portland cement component or fillers only.

8. Compressive strength—This should be directly related to concrete performance but there can be differences due to admixture interactions, different water cement ratio etc. In some countries cement is

sold as being a particular strength grade. Generally higher strength grades are more expensive but less can be used to meet a strength specification. The selection of a high strength cement becomes important when very high strength concrete is required, since an increase in cement content will not give a strength increase beyond a certain point.

It is desirable for ready-mix producers in particular to develop a good working relationship with their cement supplier. A variation-free product cannot be expected, but honesty in reporting current test results, and help in interpreting and compensating for their likely effects on concrete, and cooperation in tracking down any problems is valuable. This kind of cooperation is unlikely if all concrete problems are automatically blamed on the cement, and the concrete producer fails to carry out and keep proper records of control tests on concrete.

An important, if relatively rare, occurrence is an unfavorable interaction between the cement and admixtures in use. Examples have been encountered where a particular cement and admixture, both satisfactory with other admixtures and cements, have given trouble in combination. In one example the trouble was a false set. A false set is one that occurs for a limited time and can be overcome by continued mixing. This may give no trouble when held in a truck mixer until directly discharged into place but cause a severe loss of pumpability if discharged into a pump hopper during or prior to its occurrence. If suspected, such an occurrence can be investigated using a Proctor needle penetrometer on mortar sieved from the concrete to construct time versus penetration resistance curves.

A particularly delicate question is that of cement that provides a lower strength. It is of substantial assistance to a concrete producer if he can rely upon the cement producer advising him of a strength downturn. This enables the concrete producer to increase his cement content or make other modifications and avoid low test results. However, since the cement producer is responsible for the need for the additional cement, there is a natural tendency for the concrete producer to feel that the cement producer should bear the additional cost. It will obviously not encourage the cement producer to provide the early warning if the result is a deduction from his invoice.

The reverse kind of assistance is also valuable. Cement suppliers tend to receive unjustified complaints from customers who have inadequate control systems. It is of value to them to find a regular user who has a good control system so that they can rely on feedback data.

In summary, the development of a good relationship and an effective early warning system with your supplier can be of considerable benefit, and your own good control system is a necessary starting point for such a relationship.

2.1.3 Types of cement

Cement chemistry is extremely involved and not within the scope of this book, however limited comment on the different types of cement commonly available may be useful. All Portland cement is conveniently regarded as composed of four compounds:

C_2S, dicalcium silicate—Slow reacting, low heat generation, best long-term strength and durability

C_3S, tricalcium silicate—Quicker reacting, more heat generated, still good strength and durability but not as good as C_2S

C_3A, tricalcium aluminate—Very rapid reaction, high heat generation, responsible for early (but not high) strength and setting, readily reacts with chemicals

C_4AF, tetracalcium aluminoferrite—Relatively little influence on properties of concrete (except colour), present because needed during manufacture

The relative amounts of these compounds are varied to produce different types of cement to suit different uses:

Type I—Also known as type A, OPC (ordinary Portland cement), GP (general purpose), C

Type II—Modified low heat cement

Type III—High early strength or rapid hardening

Type IV—Low heat cement

Type V—Sulfate resisting cement

A fifth compound, $CaSO_4$ (gypsum) is interground with the cement clinker to control setting. It is also thought to have a substantial beneficial influence on shrinkage and to produce improved strength. However, an excess can cause slow setting and also unsoundness (destructive expansion). Gypsum can be rendered less effective by excessive heat during grinding.

The reader will be able to work out from the above or consult other sources about which compounds will predominate in which cements. However, there are a few matters that are often misunderstood and so should be brought to the reader's attention:

1. Sulfate resisting cement is made by limiting the amount of C_3A. Unfortunately C_3A, although responsible for expansive reactions with sulfate, happens to be the compound that also reacts with chlorides reducing their rate of penetration. In some parts of the world, this cement is assumed to be a general high-durability cement and used

where chloride resistance is as important, or even more important, than sulfate resistance, such as in marine structures. What should be used in these circumstances is blast-furnace cement, fly-ash substitution, or silica fume incorporation. Where none of these are available, an OPC concrete should be used with a low w/c ratio and possibly a corrosion inhibitor.

2. Low-heat cement is generally almost as sulfate resisting as sulfate-resisting cement (since C_3A is also limited to reduce heat generation), however, sulfate-resisting cement is not necessarily low-heat generating. This is because most of the heat generation comes from the C_3S component (of which there is always much more than the C_3A) and the proportion of this is not necessarily limited in sulfate resisting cement. Low-heat Portland cement is not readily available and has been effectively replaced by the use of fly ash or ground-granulated blast-furnace slag (GGBS) in massive elements.

It is now coming to be recognised that suitability for different purposes is often better attained by the use of variable proportions of fly ash, GGBS, natural pozzolans, or silica fume than by the use of different types of Portland cement. These alternative materials, being essentially by-products of other manufacture, used to be thought of as inferior substitutes for cement and used only to reduce cost. The reaction of concrete specifiers to these supplementary cementitious materials (SCMs) was that they were often prohibited or strictly limited in proportion.

Fortunately, the situation has changed and the benefits of SCMs are now widely accepted. For example, in the Middle East where the environment is particularly severe, many specifications require the use of SCMs to reduce heat of hydration and chloride diffusion. SCMs are not locally produced and so are imported and usually considerably more expensive than Portland cement.

2.2 FLY ASH (OR PULVERISED FUEL ASH [PFA])

2.2.1 General characteristics

Fly ash, otherwise known as pulverised fuel ash (PFA), is a pozzolanic material. This means essentially that it is capable of combining with lime (in a suitably reactive form) in the presence of water to form cementitious compounds. As lime is liberated in substantial quantities when normal cement reacts with water and is present as reactive calcium hydroxide, there is a distinct attraction in adding fly ash to concrete.

Fly ash looks like cement to the naked eye but will not set at all when mixed with water (unless a class C ash, which is a type of ash that contains substantial calcareous material). Fly ash is sometimes even finer than cement; generally spherical particle shape, including some larger hollow spheres known as a cenospheres (as opposed to the extremely jagged particle shape of cement); and is of lower density (specific gravity [SG] usually 1.9 to 2.4 compared to 3.15 for cement).

Fly ash has a varying "pozzolanicity", that is, some fly ashes give much better strength than others. No unmodified fly ash is as good as cement on a volume for volume substitution basis, but some fly ashes are as good as cement in terms of 28-day strength and better at later ages when substituted on a mass for mass basis and when account is taken of their water-reducing action as well as their strength production at a given w/c.

As fly ash is a by-product of the power industry, it can be affected by changes in electricity generation due to variation in coal source, partial use of biomass for fuel, and changes in burning temperature. As a result, fly ash is becoming a far more variable material and concrete suppliers need to be even more vigilant in assessing the quality of the material before incorporating it into their concrete. These changes in the power industry and the strong push for alternatives to power generation from coal have created an opportunity for natural pozzolans to play a more significant role in the SCM industry.

There are few materials that do not have some drawbacks and with fly-ash substitution these include

1. Reduced early strength
2. Increased setting time
3. Reduced heat generation (which is an advantage in hot weather or for mass concrete, but a disadvantage in cold)
4. Inhibition of air entrainment, if of high carbon content (easily corrected by higher dosage or specially formulated products for use with fly ash but may give rise to higher variability if carbon content varies)
5. Added complication—One more factor requiring knowledge and skill to give best results

Fly-ash concrete does not automatically display all the advantages (or disadvantages) of which it is capable. Crude substitution of fly ash for cement can yield better or worse concrete depending on the circumstances and requirements. It could be said that fly ash puts another useful tool in the hands of competent technologists and presents another trip wire for the uninitiated to fall over. Also there are considerable differences among different fly ashes and there is not an automatic best buy for all circumstances. There are examples of troubles exacerbated if not caused by fly ash and, on the other hand, of the use of fly ash not being permitted through ignorance or blind prejudice in circumstances where it would have been highly desirable.

The bridge specifications in the state of New South Wales, Australia, specifically prohibited the use of fly ash in concrete in the early 1990s. Recent surveys of the bridge stock showed the durability of the bridges built without fly ash was profoundly inferior to those with fly ash before and after the prohibition.

2.2.2 Composition of fly ash

There are two types of fly ash, according to the classification in ASTM 618, Class F and Class C. Class F ash is the true pozzolanic material, silica (as SiO_2) being the most important constituent, and alumina and iron oxide are also active (see Table 2.1). Class C ash also contains appreciable amounts of calcium compounds and may have some minor hydraulic cementing value in the absence of cement (a very few sources may produce usable concrete without any cement at all). Certainly it is possible to use it in larger proportion than Class F ash in a similar manner to, but not to the same extent as, GGBS. Class C ash may be less effective than Class F ash in providing sulfate resistance and reducing alkali silica reaction. It is possible to use high replacement levels of Type F fly ash as in the case of roller compacted concrete and massive elements. Depending on temperature, it will take a long time to develop strength.

Most of our experience is with Class F ash. Class C ash may in general produce similar effects but (as noted in the section on mix design competitions) substantial differences are possible. There is also a significant difference in the durability properties.

Carbon is the most important impurity as it can inhibit the action of admixtures, particularly air entraining admixtures. It is measured by loss on ignition, which should not exceed 6%[1] and should preferably be very much less. However, the really important requirement is that it should be as consistent as possible since otherwise it may be very difficult to control air content. However, there has been a report of rice husk ash containing up to 23% of carbon being successfully used in particular circumstances (see Section 2.5), so possibly higher percentages in fly ash would not necessarily render it useless in all circumstances.

Other impurities are alkalies and magnesium, which need to be limited in cement but are not often a problem.

Table 2.1 Typical chemical composition of cementitious and pozzolanic materials

	Portland cement	Fly ash	Fly ash (Type C)	Slag	Silica fume
SiO_2	20	50	40	35	93
Al_2O_3	5	30	15	15	2
Fe_2O_3	4	10	6	1.5	<1
CaO	65	2.5	25	40	<1
MgO	<2	<2	5	7	<1
Na_2O	<2	<2	<2	<1	<1
K_2O	<2	<2	<2	<1	<1
SO_3	<4	<3	3	<1	<1
LOI	<2	<2	<2	—	<2

[1] ASTM 618, AS 3582.

The increasing variability in fly ash may necessitate a relaxation in some compositioned components of standards. Reliance on performance specification is the way to ensure adequate concrete properties.

2.2.3 Effects of fly ash

There are three kinds of effect from the incorporation of fly ash in concrete. These are physical effects on both fresh and hardened concrete, chemical effects on setting process and hardened concrete, and physical chemistry (or surface chemistry) effects on setting process.

2.2.3.1 Physical effects

The fly-ash particles are very similar in size and shape to entrained air bubbles, and have many very similar effects:

- Water reduction—Perhaps of the order of 5% but varies with different ashes. A very few ashes (e.g., some Hong Kong ash) slightly increase water requirement.
- Reduction of bleeding.
- Improved cohesion and plasticity.
- Improved pumpability.
- Reduced slump loss with time.

Fly ash is not compressible and probably does not help frost resistance at all (and tends to inhibit air entrainment so that a larger dose of air entraining agents [AEAs] is needed). However, this property (incompressibility) makes fly ash even more valuable than entrained air for pumpability. Also fly ash has the benefit that it is present as a clearly defined quantity.

2.2.3.2 Chemical effects

When cement hydrates, it releases free lime. This lime is the softest, weakest, and most susceptible to chemical attack and leaching of all the constituents of concrete.

By reacting with the weaker and more porous portlandite, fly ash substantially reduces permeability in the hardened concrete if properly cured.

The fly ash combines chemically with the free lime to form compounds similar to those produced by the rest of the cement. This reaction occurs after cement hydration and generates less heat during hydration. This is generally a valuable property in hot climates and for mass concrete, but may be a distinct disadvantage in colder climates.

Fly ash is effectively reactive silica, the very material causing problems in aggregates through alkali–silica reaction. Actually this is a valuable feature since there is so much reactive silica that most alkali is used up during an initial reaction, leaving little to cause problems later, however reactive the aggregate.

2.2.3.3 Surface chemistry effects

It appears that fly ash can act as a catalyst or a nucleation site for crystal growth in the cement paste. Such effects are beyond the scope of this book, but it should be realised that there is more to the story than has been told here. This may provide some explanation for a smaller early age strength reduction than chemical effects alone would predict when equal mass substitutions are made.

Malcolm Dunstan in the United Kingdom and Mohan Malhotra in Canada (Malhotra and Ramezanianpour, 1994) have done interesting work on roller compacted and other concrete with 50% to 60% of fly-ash substitution. A revealing point is that good results are obtained with high-volume fly ash in dry roller compacted concrete or a low water to cementious ratio concrete using a superplasticiser. However poor results are obtained with high-volume fly ash at normal water contents (Odler, 1991). It could be said that the w/c verses strength relationship is even more marked in the case of fly ash than in the case of cement. Perhaps higher Type F fly ash replacement may be possible in low w/c concrete where full hydration will not occur. However, at higher w/c, the presence of fly ash beyond 30% reduces the amount of calcium silica hydrate produced.

Figure 2.1 shows the effect of fly ash on adiabatic temperature rise of concrete with a constant cementitious content of 400 kg/m³. At 30%

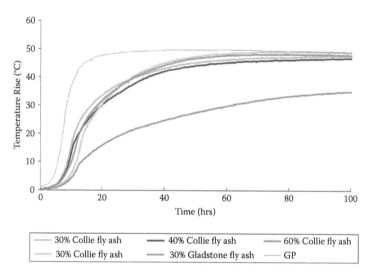

Figure 2.1 Adiabatic temperature rise curves for fly ash concretes. (From Pettinau, C. B., The Effects of the Type and Quantity of Binder on the Adiabatic Temperature Rise in Mass Concrete, final year project, Curtin University of Technology, Kent, Australia, 2003.)

replacement, fly ash was found to slow the rate of temperature rise but not change the peak temperature. At 40% replacement, the rate was slowed more and the peak temperature was reduced to some extent. At 60% replacement, both the rate and peak temperature were reduced significantly, but even after 100 hours, there was continued temperature rise. This demonstrates that a high replacement level is able to profoundly reduce temperature in mass concrete elements. As an aside, the continued temperature rise indicates that hydration processes were still continuing in spite of the considerable maturity of the concrete.

Although there are many specifications that limit the amount of fly ash replacement to 25% to 30%, replacement levels of 40% and 50% or more can be extremely useful in controlling the temperature rise in mass concrete pours. For a 3.7 m thick raft in the Middle East, the peak temperature limit could not be achieved with the specified maximum replacement level of 30% because the placement temperature was over 35°C. Redesigning the concrete mix with 55% fly ash solved the problem, all other parameters of the concrete were also achieved with the compliance age changed to 56 days. Many concrete technologists suggest that fly ash will not exhibit pozzolanic activity at replacements of more than 30%. The work on high-volume fly ash by Malhotra and others shows that it does and use of high replacement levels in appropriate applications is an important tool for the concrete technologist (Bilodeau and Malhotra, 1992). However, the use of high-volume fly ash for structural concrete should be limited to low w/cm and where good curing can be assured.

2.2.4 Advantages of fly ash

- Fly ash has reduced heat of hydration in the critical period. In the authors' opinion, up to 30% fly ash replacement, the temperature rise under adiabatic conditions is almost the same as if only Portland cement were present. The primary benefit is in reducing the rate of temperature rise allowing heat dissipation.
- Fly ash allows for more workable fresh concrete, easier to pump, compact, trowel, less bleeding and segregation usually gives a better off-form surface.
- Fly ash has substantially reduced permeability (if adequately cured).
- Fly ash allows for more durable concrete, which is more resistant to alkali silica reaction, chloride penetration and sulfate attack than concrete with Portland cements.
- Fly ash allows for higher strengths. Adding fly ash is distinctly better than using cement contents in excess of $400-450$ kg/m^3 in most cases.
- Fly ash is more economical than straight cement in most parts of the world.

- Fly ash is particularly useful in marine structures (where curing time is available before inundation) as otherwise there is the conflict of requiring high C3A to resist chlorides and low C3A to resist sulfates, whereas fly-ash concrete resists both.
- Fly-ash concrete reacts extremely well to steam curing with little or no detrimental effect unlike Portland-cement-only concrete.

2.2.5 Dangers to avoid with fly ash

- Since fly ash is lighter (and usually cheaper) than cement it might be thought that it would be especially useful in low-strength concrete. In fact it does produce much better looking concrete, which has greater segregation resistance and is less prone to bleeding for a given (relatively high) water to cementitious ratio. However this is sometimes its undoing. Uninformed or thoughtless people tend to overwater it to a greater extent than plain concrete, yet in fact its strength is more affected by a given amount of excess water. Thus fly ash should be used with care and conservatism for low strength requirements. Properly used it is valuable for such uses but is less resistant to overwatering abuse.
- Because strengths take longer to develop, more efficient and prolonged curing is necessary for fly-ash concrete. It is true that fly-ash concrete is substantially less permeable than plain concrete of similar strength, and therefore may be to some extent "self-curing" in larger masses (and especially for below ground or on ground foundations). However, this does not help the exposed cover concrete or in thin elements
- Calcium hydroxide is an end product of the reaction of C_3S and C_2S with water. The amount of calcium hydroxide in the hydrated Portland cement paste can constitute up to 26% of the total volume (Marchand et al., 2001). Although calcium hydroxide has the disadvantages of being soft, weak, and easily dissolved by water or chemicals, it is also a source of the alkalinity, which helps protect steel from corrosion. At high replacement levels, the pozzolanic reaction of fly ash with calcium hydroxide may reduce alkalinity. This is the reason that rates of carbonation can be higher in fly ash concrete and therefore the chemical protection available for the reinforcing steel. The question is whether this is compensated for by the reduced permeability of the fly-ash concrete. The answer lies in the curing: yes if well cured, no if not well cured.
- Because fly-ash concrete gains strength more slowly, it is susceptible to creep if depropped (beams and slabs) too early. The need to prop longer may be an additional cost. But remember it is the in situ maturity that counts.
- Due to reduced bleeding, plastic shrinkage cracking due to evaporation can occur more readily without appropriate precautions.

- Readiness for trowelling will be delayed, perhaps very significantly delayed in cold weather.
- The use of high-volume fly ash in massive elements will reduce peak temperature and there will generally be no detrimental effect on other properties if well cured due to the increased maturity in the large element. This is not necessarily true for smaller suspended elements that do not have the same increased maturity and are prone to greater drying due to the larger surface area to volume ratio. The tendency to specify higher volumes of fly ash to reduce Portland cement in such elements can be dangerous. Just because the 56- or 90-day strengths for the compliance specimens are adequate does not mean that the in situ strength will be in an element where typical curing may be a few days in the formwork, if you are lucky.
- Coal-fired power plants produce power not fly ash. The quality can vary significantly with changes in coal source and biomass addition. Quality control of fly ash is becoming more important.

2.2.6 Summary

The use of a proportion of fly ash is generally desirable. The exceptions are when (a) high early strength is required, (b) heat generation is advantageous, and (c) especially with strength grades below 30 MPa where adequate curing is uncertain and corrosion protection of reinforcement is required. Where fly ash is used, care must be taken to ensure that reported strengths are realistic and not the result of assuming that water-cured cylinders necessarily correctly represent poorly cured in situ concrete.

The circumstances in which it may be worthwhile specifying that fly ash be used would include hot weather concreting, large sections where low heat cement or ice might otherwise be needed, projects in which exceptionally high strength or good pumpability is needed, and projects where high sulfate resistance is needed.

2.3 SUPERFINE FLY ASH

In some parts of the world a superfine grade of fly ash is available which can be regarded as somewhere between normal fly ash and silica fume in cost, effectiveness, and desirable dose rate. The material can be highly competitive depending on relative costs and availability. It neither requires such large volume batching facilities as normal fly ash nor is as difficult a material to handle and disperse effectively as silica fume (Butler 1994).

Another important advantage of superfine fly ash is the presence of aluminate compounds which enhances the chloride binding capacity of the resultant concrete compared with pure silica based products, such as silica fume.

2.4 GROUND-GRANULATED BLAST-FURNACE SLAG (GGBS)

2.4.1 Properties of GGBS

The properties of cementitious and pozzolanic materials depend on their chemical composition, their physical state, and their fineness. This is particularly the case with blast-furnace slag. Since it is a by-product of the production of iron, its composition may differ from different sources but is likely to be reasonably consistent from a given source. Table 2.1 shows its composition to be more similar to that of cement than to typical pozzolanic materials. However, to develop satisfactory properties it is essential that the molten slag be rapidly chilled (by quenching with water) as it leaves the furnace. This causes the slag to granulate, that is, break up, into sand-sized particles. More important it causes the slag to be in a glassy or amorphous state in which it is much more reactive than if allowed to develop a crystalline state by slow cooling. In the latter state it is suitable as a concrete aggregate but not as a cementitious material. It is important to note that the unground granulated material does not make a good fine aggregate because often the grains are weak, fluffy conglomerates rather than solid particles.

To use as a cementitious material, the granulated slag must be ground as fine or finer than cement. The fineness of grind will (along with the chemical composition and extent of glassiness) determine how rapidly the slag will react in concrete.

Slag cannot be used alone to make concrete but can be used in much larger proportion than pozzolanic materials. Portland cement clinker or some other activator is required to initiate the hydration of the slag. The latter may comprise 80% or more of the total cementitious material but 50% to 65% is more usual in most parts of the world. In North America, GGBS is typically used at replacement levels of 20% to 30% similar to fly ash. However, the authors do not fully understand why North America foregoes the advantages of higher replacement levels. An alternative activator is calcium sulfate, producing a product known as "supersulfated cement". This cement is no longer produced. Although it offered the valuable properties of chemical resistance and very low heat generation, it required special care and understanding in use to offset its slow setting and strength development, and needs very thorough extended curing.

In Portland blast-furnace cement, the slag may be interground with the cement clinker or added as a separate material. The Portland cement clinker is softer than the slag and therefore will be more finely ground when the materials are interground. Even when sold as a composite "blended cement" (the term is also applied to other blends) the GGBS cement may have been either interground or postblended.

2.4.2 Properties of GGBS concrete

Concrete using GGBS cement will tend to develop early strength more slowly than pure Portland cement concrete except when very finely ground. However, if thoroughly cured, it may have as good or better eventual strength. It normally has a greater resistance to chemical attack and is particularly suitable for marine works. Its normally greater fineness may confer resistance to bleeding in the fresh state and lower permeability when hardened.

The glassy surface of the slag may give a slightly reduced water requirement even though it does not have the favorable particle shape of fly ash. The water requirement may however be substantially dependent on the fineness of grind.

It can be added as a separate ingredient at the mixer but is often sold blended with Portland cement. There is a long history of extensive use in this form as Portland blast-furnace cement, particularly in Europe and the former Soviet Union. The proportion of GGBS can exceed 80% of such cement.

To some extent this product is sometimes seen as a low-grade cement, since it develops strength more slowly and sometimes has a lower strength at 28 days. Obviously the properties of such a material will be dependent upon the composition of the particular slag. Since GGBS is a by-product material, there may be a wide variation in quality between materials from different sources. For chloride and sulfate resistance the key consideration is the aluminate component. For chloride resistance, the more the better but too much can reduce sulfate resistance. The BRE Special Digest 1 states that where the alumina content of the slag exceeds 14%, the tricalcium aluminate content of the Portland cement fraction should not exceed 10% to achieve additional sulfate resistance.

When used in lower proportion (less than 30%) in blended cement, it may be marginally cheaper and may gain strength more slowly depending on the fineness of the GGBS. It is by no means necessarily inferior.

2.4.3 Heat generation

There are three aspects to consider with heat generation. These are cold weather concreting, hot weather concreting, and mass concrete. Because it can be used in large proportion, GGBS can give rise to problems with slow setting, slow strength gain, and lack of early resistance to frost in cold weather. These same properties can be very advantageous in hot weather. The assumption may be made that the slag cement will provide reduced peak temperatures in mass concrete as does fly-ash concrete. In fact unless a high proportion of GGBS (over 60%) or a very coarse grind is used, the GGBS cement can give rise to even higher temperatures than with normal Portland cement. Figure 2.2 shows the adiabatic temperature rise for blended cements containing 25% GGBS (GB), 65% GGBS (LH), and 60% GGBS and 7% silica fume

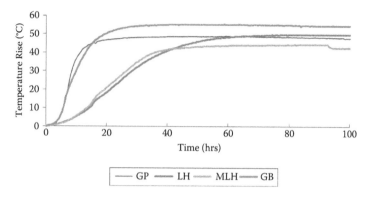

Figure 2.2 Adiabatic temperature rise in GGBS mixes. (After Pettinau, C. B., The Effects of the Type and Quantity of Binder on the Adiabatic Temperature Rise in Mass Concrete, final year project, Curtin University of Technology, Kent, Australia, 2003.)

(MLH). GGBS replacement levels at 25% (GB) reduced the rate of early temperature rise slightly but increased the peak temperature. GGBS replacement levels at 65% (LH) profoundly reduced the rate of temperature rise but did not change the peak temperature. The ternary blend with 60% GGBS and 7% silica fume (MLH) also profoundly reduced the rate of temperature rise and reduced the peak temperature. As silica fume is extremely fine and normally accelerates hydration, the reduced peak temperature is believed to be due to the silica fume reducing availability of calcium hydroxide to the GGBS.

The effect of GGBS replacement levels on the temperature rise for different concrete thicknesses is shown in Figure 2.3. The benefit of using GGBS reduces with increased section thickness. The authors have found that ponding massive concrete elements with say 75 mm (3 inches) of water is an excellent way of maximising the heat loss from massive elements as well as preventing thermal shock and providing excellent curing.

2.4.4 Blue spotting

GGBS concrete is notorious for the early development of discoloured patches, known as "blue spotting". This is caused by the initial formation of iron sulfide, which oxidises to colourless ferric salts on drying but can be a problem in continuously damp conditions or where a transparent sealer has been applied.

2.4.5 Ternary blends

Ternary (i.e., triple) blends of GGBS, fly ash, and cement are sometimes used, and have a good reputation. The addition of different proportions of fly ash during batching can give a flexibility of properties to a fixed blend of GGBS and cement.

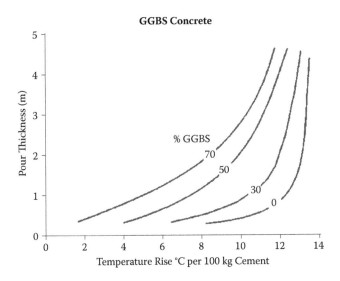

GGBS Concrete

Figure 2.3 Maximum temperature rise for different cement systems and pour thicknesses. (After Bamforth, P. B., *Proc. Inst. Civ. Engrs.*, 2, 69, 777–800, 1980.)

Nineteen-year field exposure trials showed excellent chloride penetration resistance of concretes containing GGBS. A ternary blend with GGBS and silica fume was tested and showed further reductions in chloride penetration.

2.4.6 Autogenous shrinkage

Aldred and Lim (2004) showed that low water to cementitious ratio concrete (w/cm = 0.3) containing replacement with GGBS exhibited rapid autogenous shrinkage, significantly greater than for the reference Portland cement concrete (Figure 2.4). The rapid autogenous shrinkage of GGBS concrete can be largely offset by ponding with water during the early curing period, which prevents a meniscus from forming.

2.5 SILICA FUME

Silica fume is a powerful tool at the disposal of the concrete technologist. As with other such tools, the material has to be understood and correctly used if full benefit is to be obtained and deleterious side effects avoided. In the authors' opinion, it should be used in proportions of no more than 10% of the cement content of a mix. Some specifications call for as much as 15% or more to be used which is very difficult to use, expensive, and special care is required. Because of its high surface area, silica fume should only be used together with a superplasticiser.

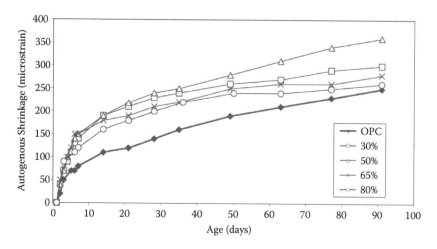

Figure 2.4 Effect of replacement percentages of GGBS on 91-day autogenous shrinkage strain. (From Alfred, J. M., and Lim, S. N., Factors Affecting the Autogenous Shrinkage of Ground Granulated Blast-Furnace Slag Concrete, 8th CANMET/ACI International Conference on Fly Ash, Silica Fume, Slag and Natural Pozzolans in Concrete, ACI SP 221, 2004, pp. 783–769.)

Silica fume (also known by one proprietary name "Microsilica") is a by-product of the manufacture of silicon, ferrosilicon, or the like, from quartz and carbon in electric arc furnaces. It is usually more than 90% pure silicon dioxide, and is a superfine material with a particle size of the order of 0.1 micron and a surface area of over 15,000 m²/kg (i.e., a hundred times greater than cement or fly ash). Its relative density is similar to that of fly ash at about 2.3 but, owing to its extreme fineness, it has a very low bulk density of only 200 to 250 kg/m³ in its loose form. For this reason it is usually handled either in a condensed form or as a 50/50 slurry with either water or a superplasticising admixture. In the condensed form, particles are agglomerated by aeration and the bulk density increases to 500 to 700 kg/m³. The problems with maintaining a stable slurry have meant that slurried silica fume is rarely used.

There is disagreement as to whether use of silica fume increases water content. This depends on how it is used. At very low water to cement ratio, the silica fume particles displace water reducing the requirement for water. To be fully effective it must be dispersed so that it occupies spaces between cement grains and must not remain in clumps of silica fume particles. Undispersed lumps of silica fume can act as sites for alkali silica reaction (Maas et al., 2007). The authors' experience and research (i.e., Lagerblad and Utkin, 1993) have shown problems with satisfactory dispersion of the silica fume. The mixing and grinding action in many mixers does not seem to be adequate to ensure dispersion. One method to overcome this tendency is to premix the densified silica fume with coarse aggregate for several minutes before batching the remainder of the concrete ingredients. It seems doubtful that this is achievable without the use of a superplasticiser and it should not be used without a superplasticiser. A possible exception may be for shotcrete, but even for this purpose the authors would normally insist on using a superplasticiser. It may be that, used with a superplasticiser, silica fume does not increase and may even reduce water content at a given superplasticiser dosage. It may also be that if any substantial increase in water requirement results, much of the potential value of the fume will be lost (especially for high-strength concrete).

There is a tendency for silica fume to be regarded as only justified for very high-strength concrete, but this is far from the truth. Its uses are many and varied. It can provide significant reductions in permeability, increased resistivity, and increased durability, and its effects on the properties of fresh concrete are more important for many uses than its effect on hardened properties. These effects include a very substantial increase in cohesion and an almost complete suppression of bleeding or any other form of water movement through concrete (in either the fresh or hardened state). Whilst the suppression of bleeding is desirable in many ways, it does cause exposed flat surfaces of fresh concrete to be very susceptible to evaporation cracking. At low replacement levels (2%–3%), silica fume is a useful pumping aid.

Some of the main applications of silica fume in concrete are discussed in the following sections.

2.5.1 High strength

The actual strength level attainable is dependent upon other factors (notably coarse aggregate characteristics) but in many instances silica fume permits the attainment of strengths in excess of 120 MPa when, for highly workable concrete, 80 MPa might be difficult to attain without it.

The action of silica fume appears to be partly chemical and partly physical. It is both superfine and in a highly reactive form. Its pozzolanic reaction with the free calcium hydroxide released by hydrating cement is therefore very effective. Ken Day has described it as being like "fly ash squared", that is, fly ash with a second order of effectiveness, for this and other properties except chloride binding due to the low aluminate component.

The physical effect of densification, and of improving the structure of the cement paste at its interface with the coarse aggregate, has been considered to be of similar magnitude to the chemical effect.

2.5.2 Durability

Silica fume concrete provides low permeability due to the chemical conversion of the most vulnerable calcium hydroxide into durable calcium silicate hydrates and improving the transition zone between the paste and the aggregate. It gives a physical uniformity of cement paste structure through avoiding bleeding effects and creating a smaller scale gel structure. Thermal stresses are reduced compared to attempting to improve durability by increased cement content.

Any tendency of the aggregate to alkali–silicate reaction will be limited since the alkalies will be consumed in a nondeleterious dispersed reaction with the silica fume.

The combined effect of these factors is to provide resistance to sulfates, chlorides, and general aggressive chemicals. Two aspects that are not necessarily greatly improved by silica fume addition are (1) carbonation and (2) resistance to freezing and thawing deterioration. In the case of carbonation, the consumption of the free calcium hydroxide in the pozzolanic reaction counteracts the beneficial effect of the reduced permeability to some extent. However, silica fume concrete has lower electrical conductivity (Vennesland, 1981), which will assist in providing greater resistance to steel corrosion.

Resistance to deterioration by freezing and thawing poses an interesting question for high-strength concrete in general. There has been debate about the presence of freezable water in the small pores in silica fume concrete. There is no question either that entrained air still provides greater resistance to freezing and thawing of saturated concrete or that it makes high strength much more difficult and expensive to attain. The question, especially with silica fume concrete, is whether laboratory tests using saturated concrete are realistic.[1] If the concrete is not saturated, there may be no water to freeze and cause damage. A different answer to this question may be appropriate in an exposed high strength column and in a bridge deck.

2.5.3 Cohesion and resistance to bleeding

These properties certainly make silica fume a most desirable ingredient of pumped concrete (and also of self-compacting concrete). A particularly severe test of pumpability occurs in stop–start situations. Many mixes pump satisfactorily on a continuous basis but fail to restart after a delay. The usual cause of this effect is internal bleeding. There is no better cure for this problem than silica fume. Using fly ash, silica fume, and a high-performance superplasticiser enabled single-stage pumping of concrete to over 600 m height on the Burj Khalifa, the world's tallest building.

Resistance to bleeding also means resistance to bleeding settlement. An important technique for very high-strength columns is to fill steel tubes from the base with fluid, self-compacting concrete. The authors have experienced this technique in four-story lifts, but there may be almost no limit to the height attainable from the viewpoint of the concrete. Such columns often

---

[1] See Section 6.3.

involve penetrations by other steelwork at each floor level. In these circumstances any bleeding settlement could be disastrous in causing cracking at vital locations.

Tremie concrete, and particularly any concrete that has to resist free falling through water, also benefits from the incorporation of silica fume, although other thickening agents such as viscosity modifying admixtures are also used.

2.5.4 Shotcrete

Silica fume concrete can transform the economics of shotcreting and greatly improve repair performance by its ability to reduce rebound and improve adherence to the substrate in both the fresh and hardened state.

2.5.5 Surface finish

The inhibition of water movement through the mix is very beneficial for surface appearance. Effects such as hydration staining, sand streaks, bleeding voids on re-entrant surfaces, and settlement cracking can be avoided. A possible problem is that the properties of the particular silica fume can cause a substantial effect on colour. This is due to any carbon content and is apparently more influenced by the size of the carbon particles than by their percentage by weight.

2.6 RICE HUSK ASH (RHA)

Rice husk ash (RHA) is produced by burning rice husks (i.e., hulls or shells) that contain a large proportion of silica. It has similarities with silica fume. Chemically it is like silica fume in being almost pure silica. Its similarity to slag is that the conditions of production are very important. As slag must be cooled very rapidly to achieve a glassy or amorphous state (glassy is amorphous as opposed to crystalline; they are not alternatives) so RHA must be burnt at between 550°C and 800°C to achieve that state. Burning at too high a temperature gives a crystalline silica that is not reactive and a health risk. However, it is important that the burning should be complete or the ash will have a high carbon and variable content, which is anathema to the uniform and effective performance of admixtures. However there has been a report of ash with up to 23% of carbon being used successfully (Dalhuisen et al., 1996). This was in tropical conditions where air entrainment was not required. Unfortunately much of the so-called RHA marketed is not true amorphous material. Most regions with rice husk resources use rice husk for energy but few have utilised the benefit of producing true RHA through controlled burning. A dangerous problem created by uncontrolled burning either in the fields or boilers is the disposal of the carcinogenic crystalline ash produced. As most of the rice-producing countries are in the developing world, the care in disposing of this dangerous waste may not always be up to the required standard. Foo and Hameed (2009) state that "the price of the ash disposal cost (either in landfills or ash ponds) hitting as high as $5/tonne in developing countries and $50/tonne in developed countries".

The particles are "fluffy". They are much larger than silica fume particles and yet have a higher surface area due to their vesicular nature. Depending on the production process, it may be necessary, and relatively easy, to grind such particles to avoid excessive water demand and resistance to compaction. With such a material, it is clearly important to evaluate product from a particular source for performance and uniformity since it can range from being as valuable as (and similar to) silica fume to being as deleterious as silt when incorporated in concrete.

There are substantial quantities (millions of tons) of rice husks available annually in many parts of the world. They constitute a potentially valuable resource if suitably prepared, rather than being a large-scale nuisance and health hazard after burning indiscriminately to reduce volume or gain energy from combustion.

2.7 NATURAL POZZOLANS

The concrete industry started with the Romans and was based on the natural pozzolans. The term *pozzolan* comes from an Italian word *pozzolana*, which means "earth of Pozzuoli", which is a city near Naples. The magnificent dome on the Parthenon in Rome is an inspiring example of the use of concrete, particularly when you consider it was built 1900 years ago. In spite of their impressive history, natural pozzolans have not been widely used in the modern concrete industry. Industrial by-products such as fly ash, GGBS, and silica fume have dominated as supplementary cementitious materials. However, the tide may be turning. Although there is still a great deal of fly ash available, modifications to coal-fired power generation have made fly ash more variable and efforts to reduce dependence on energy from coal will tend to increase this trend. GGBS and silica fume supplies are quite limited. Another important advantage of natural pozzolans as a substitute for Portland cement and other SCMs is when they are locally available and do not require importation into less developed countries.

Natural pozzolans can be classified based on their chemical or mineralogical composition or based on strength properties when it is reacted with either Portland cement or lime. For more in-depth discussion of natural pozzolans, refer to definitive guides such as ACI 232.1R-00.

Mehta (1987) classified natural pozzolans in four groups based on the principal lime-reactive constituent present: (1) unaltered volcanic glass,

(2) volcanic tuff, (3) calcined clay or shale, and (4) raw or calcined opaline silica. As volcanic tuffs generally contain both altered and unaltered siliceous glass, pozzolans of volcanic origin cannot be easily fitted into groups 1 and 2. Indeed, these are the sole or primary sources of pozzolanic activity in all groups.

The classification system adopted by the U.S. Bureau of Reclamation is based on mineral composition and shown in Table 2.2. The exact chemical composition of natural pozzolans will depend on the mineral composition of the raw or processed parent material. Metha (1987), Day (1992), and Shi (1992) summarise the chemical composition of some natural pozzolans reported in the literature.

Although most natural pozzolans contain substantial amounts of silica, additional elements are present, such as alumina and iron oxide, which also react with calcium hydroxide and alkalies (sodium and potassium) to form complex hydration compounds. This is one of the reasons why a test of the reactivity of the pozzolan is very useful, even if the chemical composition is known. It is certainly a useful quality control procedure (as it is for fly ash and silica fume). An Indian system uses the strength of lime-pozzolan mortars cured 8 days at 50°C as a means of classifying the natural pozzolan according to its strength producing properties, as shown in Table 2.3.

Natural pozzolans have been found to reduce permeability, diffusion, sulfate attack, and alkali silica reaction. However, as natural pozzolans constitute a diverse group, the required performance should be determined by appropriate testing before using. Once a particular natural pozzolan source has been selected, it would be expected to have limited variability depending on the constancy of the geological formation. Geosilica from the thermal lakes around Rotorua in New Zealand was able to be produced by blending mined material so that the final product was remarkably consistent

Table 2.2 Bureau of reclamation classification

Activity type	Essential active constituent
I	Volcanic glass
2	Opal
3a	Kaolinite-type clay
3b	Montmorilonite clay
3c	Illite type clay
3d	Mixed clay with vermiculite
3e	Palygorskite
4	Zeolite
5	Hydrated oxides of aluminium

Source: After Mielenz R. C. et al., *Econ. Geology*, 46, 3, 311–328, 1951.

Table 2.3 Indian classification of pozzolans

Activity	Strength (MPa)
Very inactive	<1.4
Inactive	1.4–2.8
Poor active	2.8–4.1
Intermediate	4.1–5.5
Active	5.5–6.9
Very active	>6.9

Source: After Hammond, A. A., Appropriate Building Materials for Low Cost Housing, *Proc. of Symp.*, Nairobi, Kenya, E. & F. N. Spon, New York, 1983, 73–83.

chemically. After processing, the benefits of this material have been found to be similar to those obtained using a high-quality silica fume.

2.8 COLLOIDAL SILICA

A French development chemically produces silica in a colloidal form rather than as a by-product from silicon ferrosilicon production. The material is even finer than silica fume but, being in a liquid suspension, does not present the same handling difficulties. It is more expensive but used at a lower dose rate than silica fume. It is claimed to be particularly effective and economical for shotcreting (Prat, 1996).

2.9 METAKAOLIN

Metakaolin is a relatively new entrant to the pozzolan for concrete field. It is produced by calcining kaolin, otherwise known as the china clay used for ceramics. As with rice husk ash, it is important that it be fully calcined but that the temperature not exceed approximately 800°C as this would cause the formation of "dead burnt", nonreactive mullite. The material is an aluminosilicate that reacts with free lime in a similar manner to silica fume and producing similar benefits when used in similar proportions of 5% to 15%.

Proponents point to the fact that metakaolin is a purpose-made controlled product, whereas most pozzolans are by-products or waste materials. Another important advantage is the relatively high aluminate component that improves chloride binding. Being essentially a white pigment, it produces concrete of a lighter shade. Since it also reduces efflorescence, it is particularly suitable for coloured concrete.

2.10 SUPERFINE CALCIUM CARBONATE (PURE LIMESTONE)

Superfine calcium carbonate (passing 75 micrometers) is another relatively recent introduction. It is primarily used in self-consolidating concrete as a fine filler instead of cementitious materials. It is usually available in varying degrees of fineness, with the superfine material being distinctly more expensive. Calcium carbonate has been used as up to 5% of OPC for many years, being seen as essentially a diluent and cost-saver. Since the material is simply calcium carbonate, it is difficult to see any chemical basis for its beneficial effects, reported to include improved workability and, more surprising, higher very early strength. The assumption is that better particle packing is at least part of the explanation as well as providing nucleation sites for hydration.

Chapter 3

Aggregates for concrete

3.1 FINE AGGREGATE (SAND)

The basic material of a natural fine aggregate is not usually a matter of concern. To some extent this has been "tested" by the formation process and any weak material broken down. There are some sands (e.g., You Yang Sand, a granitic sand from Melbourne, Australia) that are absorptive and may show some moisture movement, but generally the concerns are only with impurities, grading, particle shape, and how the sand interacts with the other materials (coarse aggregates, cementitious, water, etc.).

For too long the approach to sand quality regulation has been to consider what constitutes a "good" sand, write a specification covering these features, then accept or reject submitted sands on this basis. Sands satisfying typical specifications of this type are becoming unobtainable or uneconomic in many parts of the world, and it is necessary to devise an alternative procedure. Moreover a good sand is only good if used in the correct proportion, which is likely to differ within any reasonable specified range.

3.1.1 Manufactured sands

There have been many prejudices against manufactured sands and their use in concrete. Manufactured sands are termed many different things, for example, crusher dusts, manufactured aggregates, dust, grit, and silt. The fact of the matter is that manufactured sands do indeed differ to natural sands, and therefore need some considerations that are different to natural sands. The major differences between natural and manufactured sands are as follows:

1. They will typically have a different particle size distribution or grading (manufactured sands tend to have coarser material on the top of the grading and more finer particles).

2. Manufactured sand particles will typically have a rougher surface texture as they have recently been crushed.
3. The potential for contaminants or deleterious materials being present in the fine fraction may be higher than in natural sands due to the different processing attitude between hard rock and natural sand, and gravel producers.
4. The particle shape can be significantly more angular.

Taking these differences into account, the volume of manufactured sand percentage will start is around 5% higher than that of a good natural sand using the same coarse aggregate. So, as a starting point, rather than around 45% of the total aggregate, the manufactured sand for conventional concrete will require around 50%, and reduce through to about 40% as the powder volume in the concrete increases.

Geology does not determine how good a sand or aggregate will be. Figure 3.1 shows the compressive strength of mortars made using manufactured sands from varying rock types. A limestone yields the highest compressive strength but also the lowest compressive strength.

There is a misconception that manufactured sands have higher water demands than that of their natural sand counterparts. This may be the case in some instances, and just because it may have ultrafine particles present that does not necessarily translate into an increase in water demand.

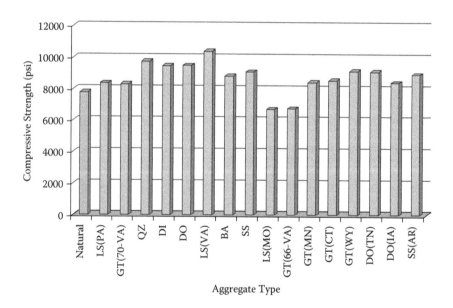

Figure 3.1 Twenty eight day compressive manufactured fine strength of mortar for different types of aggregate (fixed w/c). (From ICAR 107, 2002.)

What matters to the eventual owner of the concrete structure is not the time aggregate itself but the resulting concrete. Essentially this means that a technically satisfactory natural or manufactured sand can be defined as one that enables the production of satisfactory concrete. The required concrete properties should be fully specified by the purchaser and the sand properties should be at the discretion of the concrete producer. The same situation applies to coarse aggregates, but it is easier to justify with fine aggregates because the effects of a substandard fine aggregate tend to be more immediately experienced, especially in fresh concrete. Such effects may include retarded set, increased bleeding, excessive air entrainment, poor workability, and increased water requirement, the last leading to increased shrinkage and extra cost.

Seven features of a fine aggregate affect its suitability as a concrete aggregate:

1. Grading
2. Particle shape and surface texture
3. Clay/silt/dust content
4. Chemical impurities
5. Presence of mechanically weak particles
6. Water absorption
7. Mica content

Any of these, with the possible exception of being porous of low density, can have such serious effects on concrete as to preclude the use of the aggregate. However, this discussion will concentrate on grading, with comments on other features. This is partly because the book's views on the other six features are not significantly different to those of many others, whereas the treatment of grading is original and has permitted the use of sands considered not economically useable by others.

Much of the material in this chapter was presented in a paper titled "Marginal Sands" presented at an American Concrete Institute (ACI) convention in San Antonio, Texas, in March 1987 (and available on the website).

3.1.2 Grading

Grading is frequently regarded as the main feature of a fine aggregate, and the feature that often stops a particular sand being exploited. Although not all gradings are equally suitable for the production of concrete, there is no one ideal grading. Over the wide range of gradings that may be encountered, differences can be compensated by adjusting the percentage of fine aggregate in the aggregate combination without any need for additional cement

The basic concept is to use a smaller amount of a finer sand so as to leave unchanged both the water requirement and the cohesiveness of the mix. In any particular case, the optimum fine aggregate percentage is not solely a matter of its grading. Other factors influencing the ideal percentage include cementitious content, entrained air content, particle shape and grading of the coarse aggregate, and also the intended use of the concrete. It is assumed that the actual grading of the fine aggregate will only influence the percentage of it to be used and have no other influence on concrete properties. Although this is the case over a wide range, there must be limits to its applicability. It is necessary to be very clear where the limits are and what happens if they are exceeded.

Chapter 8 includes a thorough examination of the coarse and fine limits on the usability of a sand and on the selection of the most advantageous combination of two fine aggregates.

Grading indices

There has always been an attraction in representing a fine aggregate grading by a single number to describe its performance in concrete. This would avoid the problem of fine aggregate gradings straying into two different zones and would permit adjustment of sand percentages on a continuous scale rather than three large steps.

The original, and perhaps most widely known and used grading index is the fineness modulus (FM). This is the sum of the cumulative percentages retained on each sieve from 150 micron upward. This index is used in the ACI mix design system to adjust for sand fineness. However, it is used to indicate adjustment steps rather than to give continuous adjustment in a formula.

A problem with using a single value to express something like a grading is that a single value can represent many variations that affect performance. For example, the ACI concrete mix design method proportions the sand volume in a concrete mixture by way of the voids in the coarse aggregate and the sand FM, with FM representing a grading. However, in the majority of cases, expressing the performance of a sand by characterising it using FM alone is meaningless. Many sands will have the same FM and, as you may have experienced, will perform in different ways in the plastic concrete (see Figure 3.2).

By the same token, the coarse aggregate voids factor is influenced by several variables, including particle size distribution, particle shape, and particle surface texture. Because two void contents in coarse aggregates are the same will not result in the same performance of a concrete mix design using those two aggregates, nor will the optimal content of each of those two aggregates be the same. The same applies to particle size distribution or grading of aggregates, fine or coarse.

The specific surface (SS) is the surface area per unit solid volume (sometimes per unit mass is used). This is difficult to measure directly but may be estimated from measured or assumed values of specific surface for each individual sieve fraction in a manner similar to fineness modulus. If dealing

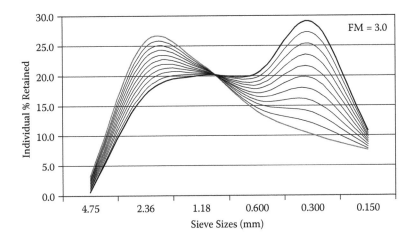

Figure 3.2 Illustration showing that multiple particle size distributions will have the same FM.

with perfect spheres, halving the diameter roughly doubles the surface area per unit weight ($1/6\,\pi d^3$ vs. πd^2). This simple assumption gives a reasonable index for aggregate proportioning. Yet what is really required is a prediction of the water requirement associated with a given amount of the fine aggregate, and cohesiveness conferred on the mixture of fine aggregate and cement paste. In general, greater surface area increases both the water requirement and the cohesiveness of the mixture. However the effect of the finer sieve fractions on water requirement is not as great as the surface area suggests (Day, 1959).

Table 3.1 (Popovics, 1982) sets out 10 factors for the numerical characterisation of individual sieve fractions. Ken Day's modified specific surface has been added to form an 11th column (the origin of Day's values is explained in Chapter 8). Some of these factors have been used as a basis for selecting the relative proportions of fine and coarse aggregates, some to calculate water requirement, and some (including Day's) for both of these purposes.

Popovics (1992) also sets out 26 formulas, 12 of which were developed by himself, for the calculation of water requirement. Some of the formulas are quite complex and tedious to evaluate, but this would be no disadvantage if the formula were included as part of a computer program. However, only a dedicated research worker could consider the time and effort that would be involved in examining the relative merits of the 26, or even the 12, formulas over a range of actual mix data.

No doubt each proponent of a system (including Day's) considers his own system quite simple to use. It is not proposed to examine all the

Table 3.1 Various proposals for sand grading indices

Limits of size fraction (columns m, e, λ, ρ, A, I, f_s)

Sieve	d (mm)	d_e (mm)	d_e (in)	S_e (m²/m³)	m	e	λ	ρ	A	I	f_s	Modified SS
3–1½ in	75–37.5	56.25 (100)	2.21 (100)	106.7 (100)	9.56 (100)	0.638 (100)	9.33 (100)	2.53 (100)	0.020 (100)	0.06 (100)	−2.5 (−100)	1
	37.5–19.0	28.25 (50.2)	1.11 (50.2)	212.4 (199)	8.56 (89.5)	1.01 (158)	11.34 (122)	3.57 (141)	0.035 (175)	0.12 (200)	−2.0 (−80)	2
	19.0–9.5	14.25 (25.3)	0.561 (25.3)	421.4 (395)	7.58 (79.3)	1.59 (250)	13.95 (150)	5.03 (199)	0.055 (275)	0.19 (317)	−1.0 (−40)	4
	9.5–4.75	7.12 (12.7)	0.280 (12.7)	842.7 (790)	6.58 (68.2)	2.53 (396)	17.49 (187)	7.12 (281)	0.075 (375)	0.27 (450)	1.0 (40)	8
No. 4–No. 8	4.75–2.36	3.56 (6.32)	0.140 (6.32)	1685 (1580)	5.58 (58.4)	4.02 (631)	22.3 (239)	10.07 (398)	0.096 (480)	0.39 (650)	4.0 (160)	15
No. 8–No. 16	2.36–1.18	1.77 (3.15)	0.0697 (3.15)	3390 (3178)	4.57 (47.8)	6.40 (1003)	29.2 (313)	14.28 (564)	0.116 (580)	0.55 (917)	7.0 (280)	27
No. 16–No. 30	1.18–0.60	0.89 (1.58)	0.0350 (1.58)	6742 (6321)	3.58 (37.4)	10.10 (1584)	39.0 (418)	20.14 (796)	0.160 (800)	0.70 (1167)	9.0 (360)	(39)
No. 30–No. 50	0.60–0.30	0.45 (0.80)	0.0177 (0.80)	13333 (12500)	2.60 (27.2)	15.94 (2500)	53.5 (573)	28.32 (1119)	0.24 (1200)	0.75 (1250)	9.0 (360)	58
No. 50–No. 100	0.30–0.15	0.225 (0.40)	0.0089 (0.40)	26667 (25000)	1.60 (16.7)	25.30 (3969)	76.8 (819)	40.06 (1583)	0.35 (1750)	0.79 (1317)	7.0 (280)	81
No. 100–pan	0.15–0	0.075 (0.13)	0.0030 (0.13)	?	0	—	?	?	?	1.0 (1667)	2.0 (80)	105

Notes: Values in parentheses are presented relative to the numerical characteristics of size fractions 3–1½ in (75–37.5 mm). d″ = average particle size; mm d_e = average particle size; z = specific surface (Edwards, 1918); m = fineness modulus; e = water requirement (Bolomey, 1947); λ = distribution number (Solvey, 1949); ρ = stiffening coefficient (Leviant, 1966); A = A value (Kluge, 1949); I = I Index (Faury, 1958); f_s = surface index (Murdock, 1960).

Source: Popovics, S., *Fundamentals of Portland Cement Concrete, Vol. 1: Fresh Concrete*, John Wiley, Hoboken, NJ, 1982.

alternatives in the current volume but, in view of the widespread use of fineness modulus, some attention should be given to it.

Table 3.2 is given in two of Popovic's books (1982, 1992) and is derived from Walker and Bartel (1947). This table provides an optimum value for the fineness modulus of the combined coarse and fine aggregates. Table 3.2 is valid for natural sand and rounded gravel having voids of 35%. Subtract 0.1 from the tabulated values for each 5% increase in voids. For air entrained concretes, add 0.1 to the tabulated values. The values are for 25 to 50 mm slump concrete; subtract 0.25 for 100 mm slump and for zero slump add 0.25.

The following equation, also from Popovics (1982), gives the water required to provide a 100 mm slump in units of pound per cubic yard (lb/cu yd) (divide by 1.685 to convert to liters per cubic metre).

$$\text{Water requirement} = c\{0.1 + 0.032[(2^m - 60)^2 + 6570]/(c - 100)\}$$

where
   m = fineness modulus of combined aggregates
   c = cement content in lb/cu yd (= kg/m^3 × 1.685)

Murdock (1960) and Hughes (1954) also introduce a term for angularity of grains. This clearly influences water requirement but cannot conveniently be used to give an adjustment to these values (see next section).

The concept of specific surface mix design is that an appropriate specific surface for the overall grading be selected allowing for the intended use. High specific surface improves the cohesion and resistance to segregation but at the expense of increased water demand. For most applications, the value is set at the lowest level that will provide a nonsegregating mix, as this gives the best economy in cement. Low workability, high-strength concrete (e.g., for precast products with heavy vibration) is resistant to segregation even with low specific surface, which reduces the water demand even beyond that which follows directly from acceptance of a stiffer cement paste. Greater workability requires more water or admixture and also a higher specific surface in order to prevent segregation. Day's mix suitability factor (MSF) summarises this concept, as shown in Chapter 8, Table 8.1.

Based on the application and recommended MSF, the sand percentage is then calculated to provide the required specific surface. The method has produced usable concrete mixes with natural sand percentages varying from 15% to 55% of total aggregates in particular circumstances, but 25% to 50% of sand is a fairly safe range.

The grading zones do not overlap because the 0.6 mm sieve is taken as the criterion. However looking at the SS values or even the FM values (Table 3.3), it is clear that the properties of the natural sands in different zones are likely to overlap. This can be avoided by defining a Zone 1 sand

Table 3.2 Optimum values of fineness modulus

Maximum size of aggregate		Weight of cement								
No.	mm	280 / 170	375 / 225	470 / 280	565 / 335	660 / 390	750 / 445	850 / 500	950 / 560	(lb/yd³) / (kg/m³)
No. 30	0.60	1.4	1.5	1.6	1.7	1.8	1.9	1.9	2.0	
No. 16	1.18	1.9	2.0	2.2	2.3	2.4	2.5	2.6	2.7	
No. 8	2.36	2.5	2.6	2.8	2.9	3.0	3.2	3.3	3.4	
No. 4	4.75	3.1	3.3	3.4	3.6	3.8	3.9	4.1	4.2	
⅜ in.	9.5	3.9	4.1	4.2	4.4	4.6	4.7	4.9	5.0	
½ in.	12.5	4.1	4.4	4.6	4.7	4.9	5.0	5.2	5.3	
¾ in.	19.0	4.6	4.8	5.0	5.2	5.4	5.5	5.7	5.8	
1 in.	25.0	4.9	5.2	5.4	5.5	5.7	5.8	6.0	6.1	
1½ in.	37.5	5.4	5.6	5.8	6.0	6.1	6.3	6.5	6.6	
2 in.	50.0	5.7	5.9	6.1	6.3	6.5	6.6	6.8	7.0	
3 in.	75.0	6.2	6.4	6.6	6.8	7.0	7.1	7.3	7.4	

Source: Popovics, S., Fundamentals of Portland Cement Concrete, Vol. 1: Fresh Concrete, John Wiley, Hoboken, NJ, 1982.

Table 3.3 Interrelationship of old UK grading zones, specific surface, and fineness modulus

Sieve size (mm)	Grading requirements (% passing)					
	Zone 1	Zone 2	Zone 3	Zone 4	AST MC33-71A	AS1465 1984
10.000	100	100	100	100	100	100
4.750	90–100	90–100	90–100	95–100	95–100	90–100
2.360	60–95	75–100	85–100	95–100	80–100	60–100
1.180	30–70	55–90	75–100	90–100	50–85	30–100
0.600	15–34	35–59	60–79	80–100	25–60	15–100
0.300	5–20	8–30	12–40	15–30	10–30	5–50
0.150	0–10	0–10	0–10	0–10	2–10	0–15
0.075	—	—	—	—	—	0–5
SS	29.40–48.31	38.54–58.31	48.00–66.00	56.00–72.00	38.00–57.90	29.40–73.10
FM	4.00–2.91	3.37–2.11	3.00–2.00	2.00–1.00	3.00–2.15	4.00–1.35
Avg. SS	38.85	48.42	52.06	63.70	47.91	41.75
Avg. FM	3.35	2.74	2.44	1.82	2.76	3.17

as a sand having an SS of 38.85 (or, say, 40, or 34 to 44); with Zone 2 being, say, 48, or 44 to 52; Zone 3 being 56, or 52 to 60; and Zone 4 being 64, or 60 to 70.

It has been contended that, to a very large extent, only the surface area and not the detailed grading of a sand is of importance. This is not completely true in all cases and the following exceptions are noted:

1. The existence of gaps in the grading (i.e., the absence of some sieve fractions) either between the fine aggregate and the coarse aggregate or within the fine aggregate grading itself can give rise to:
   a. Segregation at medium to high workability
   b. Severe bleeding
   c. Concrete that will not pump
   d. Improved workability under vibration for low slump concrete
2. Sands that are almost single-sized can give rise to poor workability through particle interference.
3. A proportion of large particles in an otherwise predominantly fine sand can cause problems through interfering with the packing of the coarse aggregate.

It is emphasised that these are rare exceptions, not glaring deficiencies in the general assumption.

Admixtures

The use of admixtures can be of considerable assistance in solving grading problems. Air entrainment is well known to have the capacity to inhibit bleeding and to assist in overcoming problems of harshness with very coarse or very angular fine aggregates. An unusual use for air entrainment is worth recounting. The mix was specified not to contain any siliceous aggregates (including natural sand) because it was to be used in the base of a furnace. This left, as the only available fine aggregate, a crusher dust with almost 20% passing a 150 μm sieve.

Day's system correctly predicted the proportion of this material that would make reasonable concrete and correctly predicted its water requirement. However, especially since a high minimum cement content was also specified, the mix was very sticky and difficult to handle from skips, even though it compacted quite well. These days a superplasticising admixture and a higher slump would probably be used, but this mix was encountered before such admixtures were readily available in Australia and in any case would have represented extra cost since the minimum allowable cement content already provided excess strength. Instead, an air-entraining agent was used and did produce a substantial improvement. It is interesting that air entrainment can both increase the cohesion of a harsh mix and lubricate a sticky mix since these are virtually positive and negative effects on the same property of the concrete. Viscosity modifying admixtures (VMA) have a similar effect to increasing the fines content. The interaction of the polymer chains can achieve similar cohesion. These admixtures are another important tool in the use of marginal aggregate gradings.

3.1.3 Particle shape

The third edition of this book strongly supported the use of natural sand over manufactured sand. However, the availability of suitable natural sands is diminishing and concerns regarding the environmental impact of their use rising. Indeed Europe is poised to outlaw the use of natural sand. We have seen that a fine sand has a higher water requirement but, over a wide range, it can simply be used in smaller proportion to give a normal water requirement. An angular sand, or especially crusher fines, also has a higher water requirement for a given grading. However, this does not justify a reduction in its proportion (it may even justify a small increase, thus further increasing water requirement, but this is fine-tuning to a precision more than required by a relatively simple system). There can be an increase in the water requirement of the mix, which may lead to an additional cost in cement or admixture when an angular fine aggregate is used. Measuring workability will indicate higher water demand. It is important to differentiate between true water demand, and admixture demand, as the latter can be a major influence.

- A coarse grade of crusher fines may be needed to fill the gap between the top of a fine sand grading and the bottom of the coarse aggregate grading. This may be essential to provide pumpability or to avoid segregation where high workability is necessary.
- It should be remembered that a higher water requirement is not purely an economic disadvantage. It also may result in increased shrinkage and so may be unacceptable for some purposes even if it is the most economical way of providing the required strength.
- There may normally be a distinct difference in colour between a crusher fines mix and a natural sand mix. One or the other may therefore be architecturally either preferred or rejected for exposed architectural concrete.
- There may be a substantial difference one way or the other (depending on actual gradings) in bleeding characteristics, which may have a substantial effect on surface appearance (coarse crusher fines being particularly susceptible to bleeding but fine dust inhibiting it).

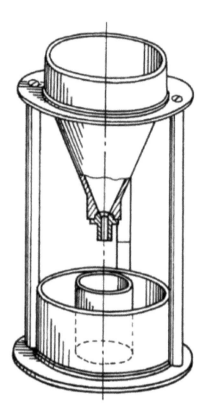

Figure 3.3 Sand flow cone apparatus.

It is often a satisfactory arrangement to use a combination of crusher fines and natural sand. Ken has formed an opinion (rather than definitely established) that there tends to be more benefit than expected from such a combination (see Figure 3.3).

Apart from gradings often fitting well together (crusher fines tending to be deficient in middle sizes and natural sand to have an excess) a small proportion of a fine, rounded, natural sand appears to have a disproportionate effect on reducing any negative effects of angularity. Also the first 2% or so by weight of silt in a fine aggregate appears not to be deleterious so that halving the amount of a silty sand will more than halve the water increasing effect of its silt.

Air entrainment and crusher fines should be approached with a little more caution. Trial mixes will very clearly show a significant advantage for air entrainment. However stone dust inhibits air entrainment and, if its proportion varies, can result in a high variability of air content, which may be unacceptable in practice. Note that fly ash (pfa) gives a similar effect on workability to that of air entrainment but is not susceptible to being inhibited or varied in its effect (other than its own inhibiting effect on air content, which is heavily dependent upon its carbon content, as measured by its loss on ignition). So crusher fines may be more acceptable in mixes containing fly ash.

The extent of the effect of particle shape can be 10%, or even more, water increase with the fine aggregate being entirely of badly shaped (but still well graded) crusher fines. However, a 7% increase is more normal for crusher fines and a badly shaped natural sand may cause as much as 3% or 4% increase. Badly shaped natural sand usually comes from glacially formed pit deposits rather than rivers or beaches. (Note that sand flow cone experimenters claim to have found fine aggregates, which increase water demand by as much as 15%.)

3.1.3.1 Fine aggregate water requirement related to percent voids and flow time

The subject of water requirement cannot be left without discussing the sand flow cone test and percentage voids. The test consists of pouring a fixed amount of dry fine aggregate into a metal funnel and allowing it to discharge into a container below, which overflows (Figure 3.3). The time taken for all the material to leave the funnel is recorded. Aggregate collected in the container is struck off to a level surface and weighed in the container. This weight, together with the container volume and dry particle density of the test material are used to calculate the percentage of voids.

Flow time and percentage of voids depend on the shape and surface texture of the fine aggregate and the grading. This is illustrated in Figure 3.4

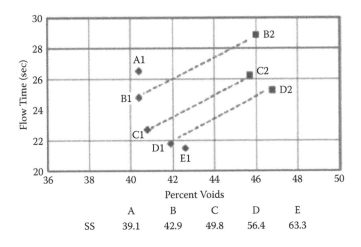

Figure 3.4 Flow test parameters of sands with controlled gradings. (From Kerrigan, B. M., Sand Flow Test, Humes Report RC 4243, 1972.)

using data from Kerrigan (1972), which shows plots of flow time and voids for sands having artificially adjusted gradings. The grading variations were applied to two basic sands to give two series: one having good particle shape and smooth surface texture (Series 1) and the other poorer particle shape and rough surface texture (Series 2). It can be seen that with deterioration of shape and surface texture, and the same specific surface (SS), the plot moves toward higher voids and longer flow time.

Norwood Harrison conducts training courses for Humes concrete technology (Harrison, 1988). As part of setting up the mortar demonstration, he obtained a quantity of a coarse but long-graded sand, and sieved it out into a series of particle sizes corresponding to each interval in the sieve mesh sizes. These were then recombined to give three standardised sand gradings: RES30 (30% passing 600 microns), RES50 (50% passing 600 microns), and RES80 (80% passing 600 microns). Sand flow results for the three sands are shown in Figure 3.5. The plot is very close to a straight line, and RES50 shows the test parameters are proportional to the change in specific surface.

Malhotra (1964) used a form of the flow test to evaluate shape and surface texture of a range of sands and the effect on workability of mortars

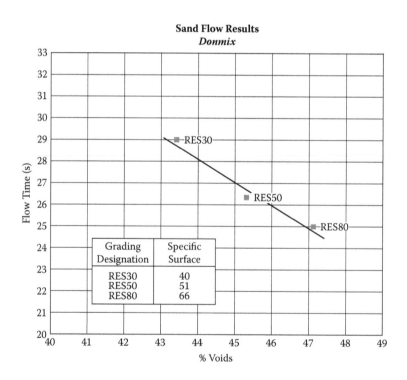

Figure 3.5 Comparison of flow time and void content for sand of varying specific surface. (From N. Harrision, personal communication, January 2013.)

made with them. The sands were sieved to provide size fractions to comply with two grading criteria and used in mortars of set composition for each of the two gradings. Workability of the mortars was assessed using a flow table. It was concluded that "the orifice test appears to be a satisfactory means of determining the shape and surface texture, and hence the water requirement, of fine aggregate".

The test has been further developed in New Zealand (Clelland, 1968; Hopkins, 1971) and independently in the United States (Gaynor, 1968; Tobin, 1978). The voids result depends little, if at all, on the dimensions of the equipment or the sample size, but different flow times will result from differences in the equipment and size of sample. It was found, for example, that even the sharpness of the transition from conical to cylindrical profile at the orifice has a marked effect on flow time (Kerrigan, 1972). Kerrigan (1972) and Elek (1973) describe a standardised test with defined sample size and dimensions of the test equipment, including the size and profile of the orifice. The specification also includes removing any particles of size greater than 4.75 mm from the test sample, as these interfere with the flow. Flow time results reported in this account of the test have all been obtained using the equipment and procedure developed and standardised by Kerrigan and Elek.

Correlation of voids in fine aggregate and corresponding water demand of concrete is acknowledged in the ACI publication "Guide for Selecting Proportions for High-Strength Concrete with Portland Cement and Fly Ash" (1998), which advises a factor of (percent voids – 35) × 8 lb/cu yd (approximately 5 kg/cu metre) amounting to approximately 15% increase in water demand per 5% increase in voids, for fine aggregates having the same grading. As the voids property of commonly used fine aggregates ranges from below 40% to approaching 48% this represents a very significant change (more than 20%) in water demand, and corresponding cement content to obtain the same performance from the concrete.

Harrison (1988) analysed data from 37 examples of concrete mixes for which both the flow test parameters of the fine aggregate and water demand of the mixes were known. The latter was expressed as a dimensionless parameter, relative water demand (RWD), being the factor between water demand of a mix made with the fine aggregate in question and a corresponding mix having fine aggregate for which voids and time plot at a particular location on a chart with axes as shown in Figure 3.6. Using linear functions, correlations were found between RWD and both percent voids and the flow time.

The results shown in Figure 3.4 and Harrison's data have subsequently been analysed further to find the positions and orientations of plane surfaces that best represent the dependence of specific surface and relative water demand separately on the flow test parameters. The outcome of this analysis is shown in Figure 3.6. Following a line of constant specific

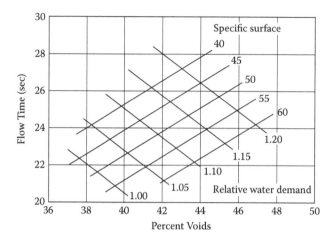

Figure 3.6 Correlation of water demand and specific surface with flow test properties.

surface we can assess the dependence of water demand on either voids or flow time. For example, for SS = 50 (a middle-of-the range value), RWDs of 1.05 and 1.20 (i.e., 15% increase) correspond to 40.2% and 45.0% voids respectively, a difference of just under 5%, very close to the estimate from the ACI parameter. The chart also shows that water demand is not linked uniquely to voids or flow time separately, but to combinations of the two properties.

The test offers a quicker and simpler means than sieve analysis of detecting changes in grading during production use of a sand. In addition it simultaneously checks for any deterioration in particle shape or surface texture. The latter may be considered fairly unlikely to change for a natural sand from a particular location but would be well worth monitoring for crusher fines and would be very difficult to check by any other means.

A further use for the sand flow cone is in blending two sands. It is a simple procedure to carry out a set of flow and voids tests with varying proportions of two sands, and a plot of the resulting properties from the flow test is very revealing as to the range of compatible proportions. An example is shown in Figure 3.7, in which the coarse sand is a low cost material, which is too coarse for use by itself in typical concrete mixes, but in blends with the more expensive fine sand gives a suitable and cost-effective fine aggregate for concrete.

In conclusion it must be emphasised that the flow test does not measure either the specific surface of a fine aggregate or its effect on water demand. Percent voids and flow time are properties that respond to characteristics of the shape and surface texture of the particles, and the grading, to which

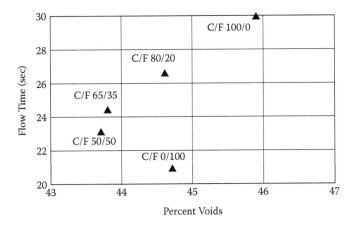

Figure 3.7 Blends of a coarse and a fine sand.

both water demand and specific surface are also related. Figure 3.6 shows "most likely" relationships based on limited data. Individual instances may not agree closely with the relationships shown, and the pattern itself can be expected to change, though perhaps not greatly, should more data become available.

3.1.4 Clay, silt, or dust content

Day's system does not provide for the incorporation of the effect of material finer than a 75 micron (200 mesh) on his "specific surface" (it is counted the same as material passing the 150 micron [100 mesh] sieve and retained on the 75 micron sieve). This is for the same reason that the effect of angular grains is not incorporated, that is, it does affect the water requirement but it does not justify an offsetting reduction in the proportion of the fine aggregate. A subsidiary reason is that the increase is not solely dependent on the weight of such material but also on its character.

It is arguable whether the 75 micron (200 mesh) sieve is worthwhile for checking fine aggregates for concrete. Certainly it is important how much of such material there is in the aggregate, but the percentage by weight gives only half the story and dry sieving rarely removes all such material. Some materials, such as the montmorillonite (smectite) clay in sand extracted in Singapore, can have three times as much effect per unit of weight as other fines such as fine crusher dust also passing the 75 micron sieve.

The definitive test for this property is undoubtedly the French "Valeur de Bleu" (Bertrandy, 1982). This test involves titrating wash water from the fines with methylene blue, which is essentially a dye composed of molecules that are single particles of standard size. The dye molecules are attracted to the surface of the fines and none remain in suspension so long as any surface area of fines remains exposed. It is possible to calculate the surface area of superfine material from the amount of the dye that has to be added before any remains in solution. This point is determined by placing one drop of the solution on a standard white blotting paper. As soon as any dye remains in solution, a faint blue halo surrounds the central muddy spot. This test is a French (tentative?) standard (also now ASTM C837-99(2003)) and is fairly easy to do in a chemical laboratory (i.e., a laboratory mechanical stirrer and a burette are needed). However, there is no point in incorporating it into Day's system because the test result would rarely be available when needed.

The alternative is very simple indeed and is the standard field settling test. Both the process of obtaining it and the use of this figure (a percentage of clay by volume when the fine aggregate is shaken up with salt solution or sodium hydroxide in a measuring cylinder and allowed to settle) are very crude indeed, but it nevertheless greatly improves the accuracy of the water prediction. The assumption made is that every 100 kg of the fine aggregate will require an extra 0.225 liters of water for each 1% by which its silt content by volume exceeds 6% (e.g., 600 kg/m^3 of fine aggregate with 8% silt content will require $6 \times 0.225 \times (8 - 6) = 2.7$ liters of extra water).

When the silt correction originated in Singapore, the sand was very coarse, requiring over 900 kg/m^3, and the silt percent was over 25% by the settling test on occasions (9% by weight). This meant that over 20 liters of additional water was required, sometimes almost 30 liters. The figure was initially derived by taking a 44 gallon drum of the dirty sand, inserting a running hose to the bottom, and overflow rinsing until the water ran clear. A repeat of the original trial mix before washing showed a water reduction of almost 30 liters. No excuse is offered for the blatant crudity of this "clay correction" because for several years now it has given good results on many different sands in Australia and Southeast Asia.

The additional water figure can be translated into an additional cement figure when the required water to cement ratio (w/c) is known. This gives a fairly precise figure for the cash value of washing the sand and so a basis for deciding whether to set up a sand washing plant. However, it is often better to counteract the effect of the clay by using a superplasticising admixture than by accepting it and using additional cement. This view has been confirmed and quantified in the laboratory by Tam (1982).

A final point on the subject of fines contents is that crusher fines dust can give a distinct (but not large) strength increase at a given w/c. In fact this is not surprising because Alexander (CSIRO Melbourne in 1950s) has shown that siliceous stone dust can have pozzolanic properties if it is ground sufficiently fine. Also calcareous stone dust (e.g., limestone) will act to some extent like superfine calcium carbonate, as discussed in Chapter 2, Section 2.10. However Day's practice is to use the settling test to allow for the extra water requirement of the fine dust but to neglect the possible strength increase.

Murdock studied the influence of fines and specific surface on the water demand on concrete, and concluded that the water demand of particle sizes that could be attributable to particle size alone disappeared for sizes lower than 100 microns.

Those who have used fly ash in concrete know that adding fine particles into a concrete mix can actually increase the workability of concrete. It is important not to confuse the change in water demand with the change in admixture demand (the amount of admixture required to get the same workability between two sands).

As each cement has its own performance characteristics, so does each sand, especially if the sand has fine materials (under 75 microns), as these small particles cease to be mere inert fillers and have the ability to influence the concrete mix performance from a chemical perspective. This is particularly true for clays.

If clays are present, these clays will have a major impact on how the concrete performs in its plastic state. This does not refer to clay lumps, but clay particles that have been either liberated from hard rock through crushing, or fine clay particles in natural sands that have not been removed by washing or classifying

Simplifying the complex chemistry between clays and the range or types of admixtures that are used in concrete, there are five types of clays that are found in aggregates.

As previously discussed, there is a misconception that when there are fine materials in sands, even clays, there will be a corresponding increase in water demand. And, not all clays are deleterious to concrete. Typically, the only deleterious clays for concrete are montmorillonite, smectite, and illite; all others are basically inert. Of course, these particles are the smallest of the small, typically a particle size less than 3 microns.

The deleterious clays actually attract admixtures to their surface, depending on the admixture and clay. Figure 3.8 shows the admixture adsorption rate at one hour with the different pure clays. This graph shows that each clay type (pure form of the clay) adsorbs different admixtures at a different rate. Some admixtures (polycarboxylate ethers, PCEs) have a higher adsorption rate than the lignin-based admixtures. This can be a major concern for using these types of water reducer compared with the older fashioned lignin admixtures if the aggregate contains clays. An alternative procedure is to use an anticlay admixture to coat the clay particles and limit adsorption problems.

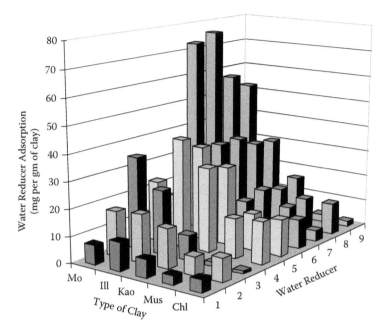

Figure 3.8 Water reducer adsorption of different proprietary products for different types of clay.

3.1.5 Chemical impurities

The question of more exotic chemical impurities is left to others, but the two questions of salt and organic impurity must be addressed. There is an extensive literature on chloride contents and their capacity to promote the corrosion of reinforcing steel. Beach sand is liable to have very high salt levels owing to the deposition of salt by evaporation. Sand dredged from the sea may be less of a problem but without washing with fresh water may still exceed a fully safe level. Salts can also cause efflorescence and higher shrinkage and affect setting and hardening rates. Although it is useful to know the chloride content of the aggregate at the source, compliance testing for chloride content should be conducted on the concrete to capture all sources of chloride. However, the limit used should be realistic so that otherwise acceptable materials are not excluded or expensive pretreatment required for no technical benefit. Another solution to chloride in the aggregate is the use of a corrosion inhibitor.

Organic impurities are quite frequently encountered in pit sands. The authors' practice is to combine the colour test (BS 812, 1960) for organic impurity with the settling test for clay content by using sodium hydroxide instead of the specified salt solution for the latter test. It is to be noted that the use of pure water will give a different result with the clay taking longer to settle and giving a higher reading. The important point to realise is that the test only establishes whether organic impurity is present and not whether it is deleterious. The colour test can fail due to the presence of a few pieces of organic matter, such as small twigs or other vegetation, that are too few and too localised to have any significant effect on strength (but could produce a visual defect on a surface).

Sands failing the colour test should then be tested for setting time and initial strength development. If they are satisfactory in these respects, it is unlikely that there will be any long-term problems (although another problem encountered has been of sands that automatically entrain air due to natural lignin).

A common effect of organic impurity (if there is any effect) is retarding or preventing chemical set. If there is no ill effect on strength up to 28 days, then the sand is satisfactory. There may be a strength reduction at 1 to 7 days but no loss of strength at 28 days, which may or may not be satisfactory for particular applications. There may be implications, with early strength loss, of setting time extension and consequent surface finishing problems for slabs.

For organic impurity evaluation, comparative mortar cubes should have the same water/cement ratio, not the same workability.

Natural impurities are not the only kind and there have been instances of accidental contamination, especially with sugar. One example was of a barge used to transport sand after transporting a load of bulk raw sugar, one result of this was to cause a large floor slab in a multistory building not to set for several days. It takes very little sugar to cause a retardation problem. In one instance a concrete strength problem later traced to employees emptying the dregs of their morning tea onto the sand pile of a small manual batching plant. They obviously liked sweet tea!

Rivaling the frequency of occurrence of all the aforementioned chemical issues with aggregates combined in the authors' experience has been the frequency of multiple dosing of retarding admixtures. This is outside the scope of the book, but it has provided more examples of concrete that has eventually proved quite satisfactory after taking several days to set. The message here is not to panic too early. If a sample sets after being in boiling water overnight (inside a plastic bag of course) then the concrete in the structure will set eventually. The question is whether it will develop serious settlement cracks in the interim due to prolonged bleeding, or to water soaking into formwork, or escaping at joints that are not watertight. It is certainly important to cover the concrete with plastic sheeting or wet hessian to stop it from drying out.

3.1.6 Weak particles and high water absorption

Weak particles and high water absorption are not common in river sands but can be encountered in pit sands. Except in very high strength concrete, or concrete required to have wear resistance or frost resistance, the direct effect on concrete strength is not likely to be a problem. Degrading during mixing, increasing fines content, and therefore increasing water requirement, is possible (but more likely in a coarse aggregate). A high water absorption may indicate an increased drying shrinkage and could also indicate a reduced freeze–thaw resistance.

3.1.7 Mica content

Except possibly in very high strength concrete, there does not appear to be a problem with moderate amounts (less than 5%) of mica directly weakening the mortar. Rather the problem appears to be an increased water requirement. Probably mica that can be seen does not do much harm, but it may indicate the presence of finer mica particles that will have much more influence on water requirement and possibly significantly increase the moisture movement tendency of the mortar.

Mica is usually detected visually but can be extracted by the use of a liquid heavier than mica but lighter than sand. However, its effect on the water requirement of mortar and therefore its strength, this time at fixed workability, is probably easier to determine and more relevant.

3.1.8 Common tests and their pitfalls

Sieve analysis/particle size distribution or grading

- Aggregates are passed through a screen with square openings. The dimension of a particle that determines the screen that it is retained on is its median dimension. However, depending on the shape of the particle, the square aperture may allow a larger particle to pass.
- Problem: Shape and size influence the result of the sieve analysis. There is also a potential error with the change in material specific gravity as particle size changes.

Fine aggregate angularity/sand flow—voids

- Aggregates flow through an orifice, into a receiving container, the volume of the container is known, and the percentage voids are calculated. This voids number is supposed to represent particle shape, surface texture, and angularity characteristics.
- Problem: The critical dimensions of the device interfere with the material flow and subsequent compaction, interfering with the voids result. Also, many combinations of the three characteristics will yield

the same voids result. The voids content by itself is not a measure of the quality of the sand. The quality (effect on water demand) is indicated by position on the plot of flow time and voids. A fine sand having high voids can be just as good as a coarser sand giving much lower voids.

Sand equivalent

- This test is used to detect the presence of clay in a fine aggregate. A solution of calcium chloride is irrigated around a volume of fine aggregate, and then the "clay" component settles out on top of the "nonclay" fine aggregate.
- Problem: The test actually identifies clay-sized particles and does not determine whether the fine material will be deleterious to the concrete mixture.

Density and absorption

- A critical test to help determine optimum material proportions, this test is limited in the size fractions that it can evaluate accurately.
- Problem: When testing ultrafine materials, the test methodology tends to skew the absorption (and hence the density) value. Also, the test is very subjective and therefore prone to operator inconsistencies.

Hydrometer analysis

- A test to determine the particle size distribution of fine materials (typically minus 200 mesh).
- The test results are calculated assuming that there is a consistent settlement rate in the test solution. The settlement rate can be impacted on by a particle's nonspherical shape and so on.

3.1.9 Concluding remarks

There is no doubt that good quality fine material can be beneficial. The questions arising are as follows: How fine? How much? In what circumstances? What is "good quality"? Hopefully this chapter provides helpful information on fineness and quantity as well as evaluating the influences of grading and particle shape, but what of the "circumstances"?

The circumstances to be considered are the content of cement, fly ash, silica fume, and so on; the properties required of the concrete, ranging from roller-compacted to self-compacting; and the presence or otherwise of air-entrainment and (especially) water-reducing admixtures.

In pure Portland cement mixes with no other material finer than 150 microns it is clear that the water requirement will be higher with a very high cement content, will reduce with reducing cement content to some optimum range (perhaps 300–350 kg/m³), and will then increase again with further cement reduction. What is happening is that, in the optimum range, cement paste fills the voids in the fine aggregate, excess cement requires additional water to form a paste with that cement, and if there is an inadequate amount of paste, additional water will be required to fill the fine aggregate voids.

Angular material in general has a higher void content than more rounded material, but the introduction of finer aggregate material, whatever its shape, may fill space that would otherwise be filled with cement paste or water. So it can be seen that this will be beneficial when cement content is below the optimum range and it should not be forgotten that cement particles are a crushed material of very poor particle shape.

Taking all this into account, there are no easy, universal answers to the question of whether a particular fine material should be used.

An excellent tool for examining the properties of fine aggregates (natural or crushed) is the New Zealand sand flow cone, as described by Harrison.

The sand flow cone is clearly suitable for examining the relative merits of different fine aggregates and different blends of two or more of such aggregates. However, it seems unlikely that it could be adapted to examination of the effects of varying cement content or, especially, the effects of superfine materials such as silica fume or of chemical admixtures. The use of rheology measurements on various proportions (after initial screening with a sand flow cone) would appear the best method to account for all components of the mix on fresh properties.

The proposed technique would not be as rapid as the dry sand flow test and would require the use of a Hobart mixer or similar, so it would probably not replace the latter.

The objective should be to establish whether optimum gradings or grading combinations established by the sand flow were still optimum under a range of contents of cement, silica fume, and other fine materials. A particularly important point would be to establish the optimum content and fineness of material passing a 150 micron sieve in manufactured sand for various types of concrete (since this is an item that could fairly readily be controlled). The test would also be useful to ensure that unfavorable reactions did not occur between cement, admixture, and superfine material (as reported in section 3.6 on mix design competitions). Perhaps small test cubes could be cast to yield a strength correction factor in mix design (i.e., to establish whether, and to what extent, the materials combination under test gave a strength increase at a given w/c ratio).

Using Day's MSF criterion, it is clear that the proportion of mortar in a cubic metre of concrete will be approximately inversely proportional to the SS of the fines used, or more strictly, to the MSF of the mortar (since the coarse aggregate makes a minor contribution to MSF). So, if the water content (Wm) of a mortar per cubic metre (of mortar) were determined, either experimentally or by a yet to be discovered calculation, that of the concrete (Wc) would be readily calculable as

$$Wc = Wm \times (\text{Required MSF of concrete} - \text{SS of coarse aggregate})/ \\ \text{MSF of mortar}$$

Interesting workability research is being done at ICAR (University of Texas at Austin) on the development and use of a highly portable rheometer. The ICAR rheometer was beta tested on the concrete for the Burj Khalifa during pumping. Sinan Erdogan investigated the effects of particle shape in both coarse and fine aggregates. The investigation is at too great a depth to present here and includes substantial work using x-ray tomography and microtomography to actually measure the shape of individual particles, in addition to using the rheometer. Very briefly he found that the particle shape of coarse aggregate does not greatly affect yield stress (which is essentially what the slump test measures) but does greatly affect the plastic viscosity (which is the part of workability the slump test does not reveal). Equally clear conclusions are not reached in respect of fine aggregates and those interested should consult the thesis (Erdogan and Fowler 2005).

3.2 COARSE AGGREGATE

The properties of a coarse aggregate depend on the properties of the basic rock, the crushing process (if crushed), and the subsequent treatment of the aggregate in terms of separation into fractions, segregation, and contamination.

Most rock has an adequate basic strength for use in most grades of concrete. Even manufactured and naturally occurring lightweight aggregates, which can be readily crushed under a shoe heel, are used to make concrete with an average strength up to 40 MPa (although they do require a higher cementitious content than dense aggregates). Exceptions to this are some sandstones, shales, and weak limestones. A different type of exception is that use involving wear and impact resistance can require a more stringent selection of rock type.

Generally, however, the stability of a coarse aggregate is more important than its strength. Rock, which exhibits moisture movement (swelling and shrinking), will add to concrete shrinkage. Again sandstone tends to be among the worst offenders, but some basalts will also display moisture movement, and some breccias or conglomerates may be quite strong mechanically and yet literally fall part after a few cycles of wetting and drying.

Rock from an untried source must be tested for susceptibility to alkali–aggregate reaction. Although comparatively rare, this reaction produces such catastrophic results that its occurrence should not be risked without at least a petrographic report. There is a rapid chemical test for reactivity but it is not very reliable. The accelerated mortar test (ASTM C1260 or C1567 for blended cement) is a better screening test. However, it will give a large percentage of false negatives. Another approach is to add sufficient pozzolans to the mix to control any possible ASR as advocated by Carse.

Another important feature of a coarse aggregate is its bond characteristics (especially in high strength concrete and where flexural or tensile strength is of special importance). This is a composite effect of its chemical nature, its surface roughness, its particle shape, its absorption, and its cleanliness. As an example of the importance of this feature is Day's experience with two different basalts in Melbourne. One of these is superior to the other on every tested feature, it is stronger, has a higher elastic modulus, is denser, has less moisture movement, and a higher abrasion resistance. However, the other aggregate was better able to produce concrete of average strength over 60 MPa. We assume that this was due to the first aggregate being so dense and impermeable that cement paste had difficulty in bonding to it. It is interesting to note that the subsequent introduction of silica fume reversed this situation, confirming the beneficial effect of silica fume on bond and the interfacial transition zone.

The particle shape of the aggregate is influenced by the crushing process. The stone type does have a distinct influence, some stones being more liable to splinter into sharp fragments or to produce a larger amount of dust than others. However, the crushing process also has a large influence. Cone crushers are perhaps the most efficient and economical type of crusher, but they do not produce as good a particle shape as a hammer mill. Other influencing factors are the reduction ratio (a large reduction in a single stage tending to produce a worse shape) and the continuity of feeding (choke feeding giving a better shape).

The effect of a poor particle shape (flaky and elongated) is to require a higher fine aggregate and water content (and therefore a higher cementitious or admixture content) for a given workability and strength. The best measure of this is the angularity number, being the percentage voids minus 33. Oddly enough Kaplan's work (1958) on the subject suggests that the sharpness of the edges and corners tends to make more difference to this parameter than flakiness and elongation.

The question of particle shape must include considering the relative merits of crushed rock and rounded river gravel. Gravels are often reputed to give inferior results, particularly for high-strength concrete. There is no denying that this is true for a given water/cement ratio and that it is true generally where tensile or flexural strength is concerned. However in terms of compressive strength, with equal cement content and equal ease of placing (reduced fine aggregate content and reduced slump [= higher yield stress] because the rounded aggregate will have a lower plastic viscosity and so can have a higher yield stress for equivalent workability) rounded gravel may give as good or better results, depending on the particular use. Forty years ago, Day made concrete of 85 to 90 MPa from London area gravel (which is one of the gravels that has been claimed to give inferior results for high-strength concrete). Gravels tend to have been adequately tested by the formation process as regards weaker particles and moisture movement susceptibility. However, this provides no security against alkali–aggregate reactivity and any coatings on pit gravels in particular should be regarded with suspicion.

The subject of coatings on coarse aggregate is worth consideration. Generally if the coating is removed during the mixing process (and assuming it to be chemically inactive) it is not likely to cause a severe problem. Very fine material will merely add to the water requirement in the same way as fine aggregate silt. This will increase water requirement but, unless excessive, should cause only a small strength depression. However, if a coating remains intact after the concrete is in place, a substantial effect on strength and durability can occur through loss of bond. The amount of fine material adhering to coarse aggregate is often substantially affected by the weather, with more material adhering during wet periods. This effect should be considered when looking for causes of strength variations in concrete.

The ideal maximum size for a coarse aggregate has usually been assumed to be 40 mm or 20 mm (1½ inch or ¾ inch) according to the size of section and the reinforcement spacing. However, there has been a worldwide trend to higher concrete strengths and work done many years ago in the United States (Blick, 1974) is gradually being rediscovered the hard way in many other places. This work showed that the optimum size of aggregates depended on the required strength level, being smaller for higher strengths. That is provided optimum is defined as that which gives the minimum cement requirement for a given strength.

If optimum is defined in terms of water/cement ratio or shrinkage or (less certainly) wear resistance, larger sizes may be best. Although the optimum size may vary from 40 mm at 20 MPa to 14 or even 10 mm at strengths over 50 MPa, the margin is not usually large and little harm is done by standardising on 20 mm. One exception to this is where difficulty is experienced in obtaining a high strength, in which case a smaller aggregate should certainly be tried. It is interesting to note that this effect has now been seen to extend further than most would have believed possible. In reactive powder "concretes" with strengths of several hundred megapascal, the coarsest aggregate used is a fine sand.

A smaller maximum aggregate may also be required for pumpability. The maximum aggregate has the greatest effect on friction factor as discussed in Chapter 8 on mix design.

Another hotly debated question is the relative merit of gap and continuous gradings. A basic difference is in segregation resistance and pumpability. High slump and pump mixes require continuous gradings but low slump, nonpump mixes compact faster with gap gradings. Two further points worth noting are that single-sized aggregates do not segregate in stockpiles and that it is more critical that the exact optimum sand percentage be used in the case of a gap grading than in the case of a continuous grading.

3.3 LIGHTWEIGHT AGGREGATES

Many types of lightweight aggregates are in use and full coverage is beyond the scope of this book. However some indication of the possibilities may be of assistance. Nonstructural lightweight concrete is not only outside the scope of the book, but also outside the scope of the mix design and quality control (QC) systems with which the book is mainly concerned. Such concretes are produced either by the use of foaming agents or the introduction of extremely lightweight aggregates such as polystyrene foam or expanded vermiculite or perlite. The range of lightweight concretes is a continuous one. It is difficult to say where nonstructural stops and structural starts. There may indeed be some overlap, with some concretes strong enough to be regarded as structural being lighter than others not having enough strength for structural purposes.

Structural lightweight concrete may be regarded as concrete having a strength at least 10 MPa and, perhaps more important, having a good degree of durability. It should also be capable of bonding to and protecting reinforcement. Such concrete is likely to have a density in the range of 1200 to 2000 kg/m³. Coarse aggregates used include naturally occurring pumice and scoria (of volcanic origin), cinders from coal burning, and manufactured aggregates produced by expanding clay or shale in rotary kilns similar to (and often formerly used as) cement kilns or air-cooled slag.

The main difficulty with lightweight aggregates is usually that they have a very high water absorption. Some aggregates, especially those manufactured in kilns, may have a relatively low permeability, sealed surface. Those that are supplied as crushed material, especially the natural materials, may absorb 20% or more of their own weight. Such materials must be used in a fully saturated state if handling difficulties are to be avoided. If this is not done, water will be absorbed during mixing, transporting, and placing, with consequent rapid loss of workability. A particular difficulty is that of pumping such concrete. Once under pressure in the pipeline, water will be forced into any unsaturated aggregate particles. This tends to cause pump blockages through severe loss of workability. The problem tends to occur on low-story work where an attempt may be made to pump concrete with aggregate that is not fully saturated. This may be successful for a limited time but as soon as any difficulty is experienced the concrete comes under greater pressure and the problem is exacerbated. Once the aggregate is fully saturated, such concrete can be pumped just as well as dense aggregate concrete. Indeed, being lighter, it may well be easier to pump to heights of 50 stories or more by reducing the hydraulic head.

It is interesting to note that at least one of the Scandinavian floating oil platforms used lightweight aggregate concrete. What was particularly interesting was that the aggregate is deliberately used dry. The Norwegians admit that this causes the problems outlined earlier but state that it was necessary to achieve the desired low density. On a dry land project, this would be ridiculous because the concrete would eventually have the same moisture content and the same density whether the aggregate was initially wet or dry. The Norwegians said that this was not the case when the concrete is to be permanently immersed in water from a relatively early age. In concrete with extremely low water to cementitious materials ratios (w/cm) the amount of water removed from saturated aggregate by hydration can help overcome this concern.

The use of saturated aggregate has benefits other than improved slump stability. The weight differential between the mortar and the aggregate is reduced, and therefore less trouble is experienced with floating aggregates. This differential is also reduced by the use of air entrainment, and the air also impedes the movement of water through the mix, so reducing slump loss. The entrapped water in lightweight concrete acts as internal curing by providing built-in reservoirs of water. This greatly improves development of hardened properties and can virtually eliminate autogenous shrinkage. Internal curing is more effective than surface curing in massive or low w/cm concrete but does not eliminate the need for surface protection. The density of the concrete is substantially affected by the moisture content and the weight loss on drying can be as much as 200 kg/m^3 with some concretes. It is also important to note that the crushing strength of the concrete may be substantially reduced by its being fully saturated at the time of test. Unlike dense aggregate concrete, lightweight concrete should not be tested fully saturated unless it will be fully saturated in use.

Lightweight concrete should not be thought of as necessarily permeable, nondurable, or less capable of protecting steel. Such material has been used to produce concrete ships and found to protect the steel very well over many years. It has been shown to give improved resistance to rain penetration in precast housing. This should not be surprising as the penetrability properties of concrete are generally determined by the properties of the matrix between the aggregate particles and the interfacial zone not the aggregate itself.

Strength capacity of different aggregates and different mixes varies considerably. Some aggregates can be used to produce concretes of 50 MPa and more, but 40 MPa is a more likely figure.

Shrinkage tends to be somewhat higher, and a higher cement content is usually needed for a given strength. These are probably both for the same reason. Lightweight aggregates will usually have a substantially lower elastic modulus and will therefore tend to shed more stress into the surrounding mortar.

The lighter kinds of lightweight concrete also use lightweight fines, but this depends substantially on the type of lightweight fines available. It is generally quite satisfactory to use any fines produced by a rotary kiln type of process, although a proportion of sand will probably be needed to give a suitable grading. However, fines produced by crushing lightweight material are often unsatisfactory. Low density is often a matter of air voids in the aggregate rather than a basic low-density material. As the material is crushed finer, more voids are exposed to penetration by the cement paste. There is a tendency to achieve little benefit in lighter concrete and a substantial disadvantage by increasing water requirement. Much structural lightweight concrete uses natural sand as the whole or part of its fine aggregate. Air entrainment often helps improve rheology as well as reduce density.

Although a slightly higher fines content may be necessary, structural lightweight concrete is generally amenable to a mix design process similar to that for normal weight concrete. Sometimes it is better to use volume batching for the lightweight material. This would apply where moisture content will vary substantially. However it is generally a matter of using the different specific gravity (SG) of the material in a similar design process. The ConAd mixtune process described in the third edition can be used for structural lightweight concrete. If so used, it is likely to require a "strength factor" of less than one. The value may be of the order of 0.7 to 0.9 but there are too many different kinds of such concrete to offer any useful guide. A trial mix will provide a factor that may prove applicable to a range of mixes using the same aggregate.

3.4 BLAST-FURNACE SLAG

The blast-furnace slag used as a concrete aggregate is quite different to the ground-granulated blast-furnace slag (GGBS) as cement. It is the same material in the molten state but has substantially different properties as a result of the cooling process. For use as an aggregate, slag must be cooled slowly to allow attainment of a crystalline state. The material is massive, requiring crushing in the same manner as a natural rock. It is also vesicular, usually to a sufficient extent to make it lighter, but not very much lighter, than a natural coarse aggregate (although it can be deliberately foamed, specifically to make a lightweight aggregate). The vesicularity means that care is needed to use the aggregate in a saturated condition if rapid slump loss and lack of pumpability are to be avoided. It also tends to cause a distinct difference in SG (particle density) between different size fractions. Excellent bond tends to be developed owing to both the vesicularity and the chemical composition of the aggregate and particle shape tends to be better than natural aggregates.

Some sources of slag may have a tendency to cause popouts as a result of remnants of crushed limestone deliberately added to provide the desired conditions in the blast furnace. However, this can be avoided if the limestone is added in smaller particle sizes and combustion is very thorough and even. Slag processing companies undertake measures to oxidise any sulfides present to prevent blue spotting. With these possible exceptions, the material tends to be a stable and satisfactory aggregate, even under fire conditions. Drying shrinkage is usually relatively low, perhaps because some chemical reaction takes place at the aggregate surface, causing a slight expansion that partially offsets drying shrinkage.

The authors have found that crusher fines produced from a particular slag source when combined with a local dune sand make a very satisfactory fine aggregate in terms of strength at a given cement content and workability, even compared to a good, long-graded, natural sand. However, it should be noted that the granulated slag, which can be ground to produce GGBS, although it may look like sand, may not perform well when so used. This is because it is in a puffed state like rice bubble cereals and so the grains are weak.

3.5 CONCRETE AGGREGATE FROM STEEL SLAG

Alex Leshchinsky

Steel furnace slag is a nonmetallic product consisting of calcium silicates and ferrites combined with fused oxides of iron (15%–25%), aluminium, calcium, magnesium, and manganese. The material, a by-product of steel manufacturing, is produced in a molten condition simultaneously with steel in a basic oxygen furnace. After the air-cooling, the material has a predominantly crystalline structure. Air-cooled steel slag is crushed and screened for the aggregate.

Steel slag aggregate is being used in asphalt and road base. In asphalt, replacing natural aggregate with steel slag aggregate brings some advantages, such as improvement in skid resistance and enhancement in durability. However, the demand for steel aggregate is much lower than its output from steel operations. Therefore, steel slag aggregate is usually very cheap. The average world market price for steel slag aggregate is of the order of US$0.50/t.

The surplus of this cheap material has led to attempts to accommodate it in concrete. Maslehuddin et al. (1999), conducted detailed research of steel slag as concrete aggregate. They investigated compressive and flexural strength, water absorption, drying shrinkage and other properties of concrete. Steel slag aggregate used in the experiments contained clay lumps and friable particles in the range of 0.07% to 0.31%. Concrete with coarse aggregate from steel slag has been assessed against concrete with limestone aggregate. On the basis of the results of the study, its authors concluded that steel slag aggregate can be beneficially utilised in Portland cement concrete but highlighted concerns with possible durability problems caused by the lime expansion and aesthetic problems associated with the rust on the surfaces.

Steel slag aggregate is a very abrasive material and will result in substantial wearing of plant equipment (conveyer belts and bins) as well as agitators. Due to the high density of steel slag (an apparent particle density of the order of 3.3 t/m^3), concrete density will increase making it suitable for applications requiring high density concrete. For instance, concrete with 1 t/m^3 of crushed river gravel (an apparent particle density of 2.65 t/m^3) has a density of 2.44 t/m^3 and is delivered in maximum size loads of 6 m^3. If crushed river gravel is replaced with steel slag aggregate, the maximum load size will be only 5.45 m^3, which will increase concrete transportation cost.

3.6 CONCLUSION

As is the primary theme of this edition, it is concrete performance not component performance that is important. Consideration of aggregate quality, other than its durability, should be the concern of the premix supplier not the consultant or the regulator. Removing prescriptive components for aggregate from concrete specifications is a most important requirement for improving the sustainability of concrete.

Chapter 4

Chemical admixtures

Since the third edition of this book, no area of concrete technology has seen greater change than chemical admixtures. These advances in concrete admixtures have facilitated the use of concrete in ever-increasing applications. Table 4.1 from the "Report on Chemical Admixtures" (ACI 212.3R-10) from the American Concrete Institute summarises the vast array of materials available to change the fresh or hardened properties of concrete. Indeed, manufacturers are now able to specifically modify their polymers within these generic groups to further modify certain properties. On the Burj Khalifa project, the technical requirements for concrete included adequate retardation and workability retention for single-stage pumping to 600 m with ambient temperatures up to 50°C as well as achieving over 10 MPa compressive strength at 12 hours and 80 MPa at 28 days. The admixture supplier modified an existing product to achieve the required performance. This "just-in-time" admixture development is far removed from the tortuous testing procedures that used to be required to get a product approved for use. Up to the 1980s, many specifications excluded the use of admixtures, which was clearly not the best way to deal with what was to become one of the most important methods of modifying concrete properties. However, some caution is warranted when using complex chemicals in a very complex chemical system such as concrete. Many of the materials problems with concrete have occurred because of an inadequate appreciation of the interaction of different factors on concrete properties. For example, the early promotion of superplasticisers for flowing concrete sometimes failed to account for the limited ability to control water content in general concreting.

As set out on his website (http://www.kenday.id.au), Ken Day experienced a situation in which his mix submitted as a competition entry actually completely failed to set at all. The cause was a complex interaction of the admixture, the particular cement, and a large proportion of Type C fly ash. The effect was predictable by the most senior researcher of the admixture supplier but unknown to senior company technical representatives in both Australia and the United States. The product is described on the web

Table 4.1 Admixtures, their characteristics, and usage

Admixture type	Effects and benefits	Materials
Air entraining (ASTM C260 and AASHTO M154)	Improve durability in freezing and thawing, deicer, sulfate, and alkali-reactive environments. Improve workability.	Salts of wood resins, some synthetic detergents, salts of sulfonated lignin, salts of petroleum acids, salts of proteinaceous material, fatty and resinous acids and their salts, tall oils and gum rosin salts, alkylbenzene sulfonates, salts of sulfonated hydrocarbons.
Accelerating (ASTM C494/C494M and AASHTO M194, Type C or E)	Accelerate setting and early-strength development.	Calcium chloride (ASTM D98 and AASHTO M144), triethanolamine, sodium thiocyanate, sodium/ calcium formate, sodium/ calcium nitrate, aluminates, silicates.
Water reducing (ASTM C494/C494M and AASHTO M194, Type A)	Reduce water content at least 5%.	Lignosulfonic acids and their salts. Hydroxylated carboxylic acids and their salts. Polysaccharides, melamine polycondensation products, naphthalene polycondensation products, and polycarboxylates.
Water-reducing and set-retarding (ASTM C494/C494M and AASHTO M194, Type D)	Reduce water content at least 5%. Delay set time.	See water reducer, Type A (retarding component is added).
High-range water reducing (ASTM C494/ C494M and AASHTO M194, Type F or G)	Reduce water content by at least 12% to 40%, increase slump, decrease placing time, increase flowability of concrete, used in self-consolidating concrete (SCC).	Melamine sulfonate polycondensation products, naphthalene sulfonate polycondensation products, and polycarboxylates.
Mid-range water reducing (ASTM C494/ C494M, Type A)	Reduce water content by between 5% and 10% without retardation of initial set.	Lignosulfonic acids and their salts. Polycarboxylates.

Table 4.1 (Continued) Admixtures, their characteristics, and usage

Admixture type	Effects and benefits	Materials
Extended set control (hydration control) (ASTM C494/C494M, Type B or D)	Used to stop or severely retard the cement hydration process. Often used in wash water and in returned concrete for reuse and can provide medium-to-long-term set retardation for long hauls. Retain slump life in a more consistent manner than normal retarding admixtures.	Carboxylic acids. Phosphorus-containing organic acid salts.
Shrinkage reducing	Reduce drying shrinkage. Reductions of 30% to 50% can be achieved.	Polyoxyalkylene alkyl ether. Propylene glycol.
Corrosion inhibiting (ASTM C1582/C1582M)	Significantly reduce the rate of steel corrosion and extend the time for onset of corrosion.	Amine carboxylates aminoester organic emulsion, calcium nitrite, organic alkyidicarboxylic. Chromates, phosphates, hypohosphites, alkalies, and fluorides.
Lithium admixtures to reduce deleterious expansions from alkali–silica reaction	Minimise deleterious expansion from alkali–silica reaction.	Lithium nitrate, lithium carbonate, lithium hydroxide, and lithium nitrite.
Permeability-reducing admixture: nonhydrostatic conditions (PRAN)	Water-repellent surface, reduced water absorption.	Long-chain fatty acid derivatives (stearic, oleic, caprylic), soaps and oils (tallows, soya based), petroleum derivatives (mineral oil, paraffin, bitumen emulsions), and fine particle fillers (silicates, bentonite, talc).
Permeability-reducing admixture: hydrostatic conditions (PRAH)	Reduces permeability, increased resistance to water penetration under pressure.	Crystalline hydrophilic polymers (latex, water-soluble, or liquid polymer).
Bonding	Increase bond strength.	Polyvinyl chloride, polyvinyl acetate, acrylics, and butadiene-styrene copolymers.
Colouring	Coloured concrete	Carbon black, iron oxide, phthalocyanine, raw burnt umber, chromium oxide, and titanium dioxide.

(Continued)

Table 4.1 (Continued) Admixtures, their characteristics, and usage

Admixture type	Effects and benefits	Materials
Flocculating	Increase interparticle attraction to allow paste to behave as one large flock.	Vinyl acetate-maleic anhydride copolymer.
Fungicidal, cermicidal, insecticidal	Inhibit or control bacterial, fungal, and insecticidal growth.	Polyhalogenated phenols, emulsion, and copper compounds.
Rheology/viscosity modifying	Modify the rheological properties of plastic concrete.	Polyethylene oxides, cellulose ethers (HEC, HPMC), alginates (from seaweed), natural and synthetic gums, and polyacrylamides or polyvinyl alcohol.
Air detraining	Reduce air in concrete mixtures, cement slurries, and other cementing applications.	Tributyl phosphate, dibutyl phosphate, dibutylphthalate, polydimethylsiloxane, dodecyl (lauryl) alcohol, octyl alcohol, polypropylene glycols, water-soluble esters of carbonic and boric acids, and lower sulfonate oils.

Source: American Concrete Institute (ACI) Committee 212, ACI 212.3R-10, Report on Chemical Admixtures for Concrete, 2010.

as "especially suitable for use with fly ash mixes" without any warning as to type and proportion of the latter. The admixture in question was Grace WRDA, but it is emphasised that this admixture is a very normal lignosulphonate that has been in wide use in many countries for many years. It seems that the same effect might have occurred with other similar competing products. The point in relating this incident is that, until its occurrence, Day had for many years been happy to design concrete mixes over the telephone in many countries, and recommended that the first trial mix be a full size delivery to the actual structure, without encountering any problem. He has also recommended readers to find and rely on the technical representative of a reputable admixture supplier. Clearly this advice must now change, and concrete producers, while still listening to advice, must satisfy themselves through trial mixes before believing it.

The authors have seen trial mixes undertaken with the "same" admixture, which had profoundly different performance in concrete with otherwise the same mix composition. This was clearly due to a different chemistry where the supplier was using the project as a product development laboratory. This sometimes cavalier attitude to modifying admixtures does involve some danger because the modifications may result in an unforeseen incompatibility in the concrete. One of the reasons for ongoing modifications of admixtures without changing the product name has been to avoid the requirement for additional standard or compliance testing. The test requirements to comply with ASTM C494, for example, involve extensive testing of up to one year, especially if freeze–thaw testing is required. Recently AS 1478 removed the requirement for one-year compressive strength testing, but the range of test is still extensive and testing drying shrinkage is expensive and time consuming. The standards also suffer from nominated test mixtures that may be profoundly different from the actual application. It is no wonder that suppliers would want to avoid unnecessarily conducting such tests, but there is the danger of a significant problem occurring because of an unforeseen incompatibility or negative reaction. If it is accepted that trial mixes may be inaccurate and that other user's production results may not be applicable, the only remaining practical selection basis is an extended parallel trial. This may be simply a matter of using the admixture on trial in one or two trucks per day and always testing these trucks. Over a period it will be accurately seen whether there is any significant advantage from using the new admixture. It may be considered necessary, for a short initial period, to supply the special trucks to a noncritical location or for a use for which a lower grade has been specified. There are many cubic metres of blinding concrete for different projects on which we have worked that have much higher quality control testing and performance than anticipated in the specification. There is nothing like building up data on delivered concrete for noncritical locations to give one confidence that the performance is adequate.

The problem of optimising and adjusting mixes is further complicated by requirements for the approval of mix designs by statutory bodies and specifiers. Theoretically any change in mix proportions or ingredients may require additional trial mixes and testing. Sometimes this includes chloride diffusion limits, which can take months to measure. All parties involved hope that the proposed mix passes or the whole process will need to be repeated. Accordingly, the current system, which attempts to keep control over deviation, actually prevents appropriate modification of aggregate proportions to maintain the grading curve as discussed in the quality control section (Chapter 10) and the use of newer admixture technology that may have significant advantages to avoid the painful testing protocol. In Australia, which has a proud heritage in the concrete industry, there are many examples of obsolete admixtures and inefficient mix designs being used to avoid going through the bureaucracy of getting more suitable admixtures and mixes approved.

In presenting the theme report on production of HSC/HPC at BHP 96, the Paris symposium, Day (1996) remarked that of the more than 20 submitted papers included in his report only 1 specifically dealt with a superplasticiser, but all the concrete covered by the reports contained a superplasticiser. There may be a temptation to think that the use of silica fume, or high strength, is the outstanding characteristic of high-performance concrete but probably its most basic and essential feature is the use of a superplasticiser or high range water reducer (HRWR).

Admixture technology is both extensive and virtually a foreign language to many in the concrete industry and related professions. It is easy to provide more detail than can reasonably be absorbed and retained by non-specialists. This chapter is therefore aimed at providing general guidance rather than at providing detailed knowledge. What is new is that the situation has now become so complex that even the technical representatives of major admixture suppliers do not have all the answers.

It is important to realise both the complexity of the situation and the inaccuracies inherent in any attempt to compare the relative value of different admixtures. Different admixtures can have significantly different relative benefit when used with different cementitious materials or other different conditions. A particular brand name of admixture may be differently formulated in different parts of the world. A difference in the time of addition (relative to that of the cement first coming into contact with the water) can substantially affect the performance of an admixture. Different results may be obtained from the same mix and admixtures when mixed in a laboratory mixer or in a truck.

The basic cost of most admixture raw materials is relatively low compared to the selling price of the admixture. This is at least partly due to the very considerable costs of research and development, quality control, technical service, and marketing. However, with the possible exception of very large concrete producers with good facilities and very knowledgeable staff, the availability of technical assistance from an admixture supplier may be good value for the money.

If one admixture enables the saving of 5 kg of cement per cubic metre of concrete more than another, this may save several hundred tons of cement per annum. However, the strength difference at the same cement content would only be of the order of 1 MPa and this may be within the margin of error of the trial mixes used. There has been a worrying trend to use much higher cementitious contents than necessary and not to use admixtures to reduce cementitious contents. Specifications that require minimum cementitious contents are part of the reason but ignorance or laziness by premix suppliers is also a component. Working toward the optimum minimum cementitious content should be the aim of everyone in the industry in this era where sustainability is becoming more important.

On the whole it is probably of greater importance to select the correct type of admixture and to use it in the most advantageous way than to obtain the most cost-effective admixture. It is therefore again emphasised that most concrete producers should be seeking the ideal admixture supplier rather than the ideal admixture because the correct advice may be more important than the best admixture.

4.1 SPECIFYING ADMIXTURE USAGE

Concrete users should avoid specifying the use of particular admixtures unless absolutely essential for a particular purpose. If they do so, it should only be after the premix supplier has satisfied itself that the other performance properties are not affected, otherwise the responsibility of the concrete supplier for the performance of the concrete will be substantially reduced and any and every problem encountered will in some way be blamed on the specified admixture. As far as possible the concrete supplier must be left to formulate its concrete and this should include the use of its choice of admixtures. Where a particular admixture is considered essential, this should be discussed with the concrete supplier and an attempt made to have him use it of his own volition. If it became normal to impose the concrete user's choice of admixture on the concrete producer, this would sabotage his entire control system, as results could not be grouped together for analysis.

As with other aspects of mix design, the purchaser should be entitled to know what is being used in his concrete and to have the right of objecting to unsatisfactory proposals. In general, this right should not be used lightly. The purchaser should certainly refuse permission to use admixtures containing any significant amount of chloride in concrete to contain reinforcement or water resisting admixtures or durability enhancing admixtures with no history. This is because unsatisfactory long-term performance may result.

Where resistance to freezing and thawing or salt scaling is required, the purchaser should certainly specify that air entrainment be provided. It may also be reasonable to object to an air entrainer that produces too large a bubble size and has an unsatisfactory spacing factor. This is because it is the spacing of the air bubbles that matters for frost resistance, whereas the total volume is what is measured by all typical tests and what affects the strength of the concrete. Until recently the spacing could only be determined by microscopic examination of a cut-and-polished face of hardened concrete. Now the air void analyser (AVA) enables air void parameters to be measured in 25 minutes or less on fresh concrete.

The AVA is a great piece of equipment but not all sites will have one and therefore air content will be the most common method used. Research by U.S. Federal Highway Administration (FHWA) shows that there can be significant differences in freeze–thaw resistance depending on the air-entraining admixture used.

4.2 POSSIBLE REASONS FOR USING AN ADMIXTURE

Reasons to use an admixture are as follows:

1. To save money and reduce peak temperature by reducing cementitious content for a given strength and workability
2. To improve concrete properties, including:
   a. Reduction of bleeding or segregation
   b. Compensation for aggregate grading deficiencies
   c. Reduced permeability
   d. Improved pumpability
   e. Reduced shrinkage
   f. Improve durability
3. To compensate for weather conditions or haulage distance, for example, retarders and accelerators
4. To reduce labor costs—Superplasticisers/HRWRs
5. To produce self-compacting concrete to facilitate placement in difficult locations, provide good off-form finish and reduce labor
6. To facilitate the use of marginal cementitious materials or aggregates.

4.3 TYPES OF ADMIXTURES AVAILABLE

4.3.1 Water reducers

The most common water reducers are lignosulphonates, which are natural retarders but may be modified by the addition of accelerators such as triethanolamine (hopefully no longer calcium chloride as in the past).

A water reduction of the order of 5% to 10% is obtained and the admixture is used basically to enable cement reduction. Some of the water reduction is due to the entrainment of 1.5% to 2% of air by this type of admixture. When an accelerator is used to reduce retardation, it can cause an increase in shrinkage but this is offset to some extent by the water reduction. There is some evidence that early shrinkage is less compensated than later shrinkage and this may lead to slightly increased susceptibility to early cracking.

The time of addition of these admixtures may be important, a delayed addition giving substantially more effect. In some cases readiness for trowelling of slabs may be delayed even when the 24-hour compressive strength is not reduced.

Water-reducing strength increasers containing polymers such as hydroxycarboxylic acids and polysaccharides can be very similar to lignosulphonates. The cement saving is of a similar order but the action is a little different since water reduction is slightly less and there is a small direct strength increase at a given water/cement ratio. These admixtures may be a little more effective in cement saving than lignosulphonates (especially at higher cement contents) but are more sensitive to variations in cement characteristics. Newer types of admixture (described as "synergised" by some manufacturers) often combine polymers and lignosulphonates in an attempt to get the best of both characteristics.

4.3.2 Superplasticisers or high range water reducers (HRWRs)

HRWRs have become distinctly more important in the years since the first edition of this book. It is hard to imagine a high-performance concrete (HPC) without an HRWR. Their wider use and greater importance have been accompanied by a better understanding of their strengths and weaknesses. It is becoming apparent that denser packing of the paste fraction of concrete is the key to higher strength, reduced permeability, and so on. This can be achieved by the use of finer materials such as silica fume, finer cement, and superfine fly ash. Such finer materials have a higher water requirement, which can offset their benefit. The answer to this is to use the fine material together with an HRWR to counter the higher water requirement. It has also become apparent that not all HRWRs are compatible with all cements and cementitious materials. The best way to check on this is to use the admixture at the intended dose in an otherwise normal Vicat setting test. Better still the test can be repeated at different dosage rates to establish the saturation dosage (i.e., that dosage above which no further water reduction is obtained) as well as checking on the possible rapid workability loss which is the nature of the incompatibility of some admixtures and cements. Alternatively it may be found that excessive retardation of set is experienced in some cases. It is also desirable to include in this test any pozzolanic materials intended for use in the concrete.

The original superplasticisers were melamine formaldehyde and sulphonated naphthalene. The former originated in Germany and the latter in Japan. These are effective water reducers with a limited period of effectiveness and apparently no significant detrimental effects on retardation or air entrainment. They are relatively expensive and use in higher volume compared to normal water reducers and cannot be justified on cement reduction grounds for ordinary concrete but usually can be for concrete with higher performance requirements.

They can be used in four ways:

1. To produce "flowing" concrete—Such concrete can be virtually self-compacting if appropriate modifications to the mix are made and may be justified on labor-saving grounds. It may also be worthwhile where excellent surface finish (on vertical formed surfaces) is required or for very congested sections (see also Section 4.3.9).
2. To produce very high strength or durability—At normal workability the water reduction can give high strength increases. This may only

be financially worthwhile when the strength required cannot be obtained by increased cement contentitious. On the other hand a superplasticiser is very desirable with high cementitious content, as the cement may not otherwise be adequately dispersed.

3. To limit shrinkage—In thin walls with congested reinforcement a small aggregate, high slump mix may be necessary to achieve full compaction. Such concrete would have excessive shrinkage if the high workability were attained by increased water and cement content, but not if obtained by using a superplasticiser at normal water and cement contents.

4. To limit cementitious content—In massive elements where excessive temperature rise could occur or where sustainability concerns warrant a reduction in total cementitious content.

These remarks apply to what are now described as first-generation superplasticisers. Nowadays, melamine formaldehyde has virtually disappeared from the market. The situation has now become much more complicated in that there are second- and third-generation HRWRs that retain their action over a considerable period of time (in some cases more than 2 hours).

The original materials derived their effectiveness not so much from a new property as from an absence of two old properties. They can be used at much higher dose rates than normal water reducers because they do not either retard set or entrain air. As an example of this, it was required to produce a highly fluid mortar with a very low water to cement ration (w/c) to surround and protect a steel tension pile (or ground anchor). High strength was really only essential at the rock anchorage over 30 m below ground level. A superplasticiser was considered, but it was realised that a normal water reducer at the same dosage would produce a similar water reduction at lower cost. It was an advantage that a very long retardation resulted (because the mortar was placed first and the pile was lowered into it). The high air percentage was reduced to a very modest amount by the fluid pressure at the full depth.

There is now an enormous variety of HRWRs available, from a dozen or more different countries. The original materials have been supplemented or replaced by others, including lignosulphonates formulated to entrain reduced amounts or air and produce less retardation. Their cost, relative to the cost of labor, is reducing. The value of very high-strength concrete is becoming more widely realised. Perhaps more important still, it is being realised that these materials are not only labor-content reducers but also skill-requirement reducers. For all these reasons, the use of superplasticisers is on the increase.

The new kid on the block is polycarboxylate ether (PCE). These admixtures are particularly favored for use in self-compacting concrete (SCC), having longer workability retention with less set retardation and apparently giving some bleeding resistance. A problem with this type of admixture is that it tends to entrain more air, which is countered by the inclusion of a defoaming agent (i.e., air entrainment suppressor). However, some such combinations require continuous agitation to avoid settling out. It also tends to be more sensitive to cement type and can result in greater workability loss at intermediate workability.

PCE and naphthalene sulphonate admixtures are totally incompatible with each other. Therefore, the premix supplier should use the same type of admixture for all concrete on a particular project. Incompatibility may also effect production if both admixture types have a common delivery path.

4.3.3 Retarders

Set retardation to any desired extent is readily available with no deleterious effects, with or without water reduction.

Sugar is a powerful retarder and very small quantities can produce a dramatic effect. It should be noted that set retardation is not the same thing as workability retention. Mixes containing water-reducing retarders may lose slump more rapidly than plain concrete in some circumstances.

Delayed addition may be very important because a greater effect is obtained by a delay of the order of 5 minutes after the water has been in contact with the cement. When retarding admixtures are added with the mixing water, the retarder can retard the release of gypsum, which is added to cement during manufacture to control rapid setting resulting a more rapid set. It is not usually practicable to actually delay addition in ready-mix operations, but the same effect may be obtained if the undiluted admixture is added at the end of the batching process and takes some time to disperse through the mix. This problem appears to have diminished and most producers add retarding admixtures to the mixing water with apparent impunity.

4.3.4 Accelerators

Set acceleration, unlike retardation, is only obtainable within limits and with some risk (or certainty) of deleterious side effects. Most accelerators tend to increase shrinkage. The trend is for high early strength to be achieved by the use of nonretarding HRWR in many applications rather than chemical accelerators to avoid any detrimental effect. The field of accelerators in particular is one in which development work is occurring and details are not readily available. The information given next is likely to prove outdated. Purchasers will need to carry out their own trials.

Calcium chloride is by far the most economical and effective accelerator. However, it has the severe disadvantage that it promotes the corrosion of reinforcement (and any other embedded steel), particularly if not properly dispersed. Many, but not quite all, authorities claim that it also increases shrinkage quite substantially. Calcium formate and calcium nitrite, which is primarily used as a corrosion inhibitor, produce almost similar strength gains but less effect on setting times. Both are substantially more expensive than calcium chloride. Sodium silicate and aluminate as well as sodium or potassium carbonates are powerful set accelerators but reduce strength at later ages. Triethanolamine and salicylic acid are only mild accelerators and are not used alone.

Hot mixing water or steam curing can also be used to accelerate set and strength gain. Hot water is in fact often a quite suitable choice as an accelerator, especially in cold climates. A major project involving thousands of very large precast segments for an elevated roadway again demonstrated this. Faced with a requirement to attain 18 MPa in 7 hours, only 2 weeks were available to solve the problem. It took only a theoretical analysis and two sets of four trial mixes each to convince the client that hot mixing water was a more economical solution than steam curing, chemical accelerators, or extra cement. The point is, given the very short curing period, that hot mixing water takes immediate effect, whereas steam curing has to be gradually applied. Of course, a superplasticiser was also used and Day's early age system (see Chapter 7) was an integral part of the solution. Insulation also plays an important role when using hot water for acceleration.

Superplasticisers are very useful for high early strengths, because they enable low water/cement ratios, which not only increase eventual strength but also increase the proportion of that strength developed at earlier ages. Also they give a strong dispersing effect, which makes more effective use of high cement contents. Some producers, particularly in tropical climates, find that using a superplasticiser is an economical substitute for steam curing precast units. Of course, such a substitution provides a very large strength margin at later ages. The continued development of superplasticisers coupled with hot water and insulation will probably completely replace traditional accelerators.

An important recent development is a patented product called x-seed. This contains particles of calcium silicate hydrate (CSH), which act as nucleation sites for further CSH formation. The result is reduced dormant phase and strength increase similar to steam curing. The resultant increase in strength appears to be without the detrimental effect of traditional accelerators on shrinkage.

4.3.5 Air entrainers

It is of interest that most concrete of up to 30 MPa (4,500 psi) in Australia contains entrained air but the practice appears unusual in Southeast Asia and Europe. Worldwide, one of the principal benefits of air entrainment is greatly enhanced resistance to damage by freezing and thawing, but in Australia, as in Southeast Asia, this is not a problem.

The other reasons for using air entrainment are

1. Reduced bleeding
2. Improved cohesion
3. Grading rectification

4. Reduced penetrability
5. Improved pumpability for short pump distance (but high air content decreases pumpability)
6. Better surface finish

The amount of entrained air required for these purposes is somewhat less than may be required for high frost resistance, 3% to 4% being normal in Australia. The disadvantage of air entrainment is that it is an additional factor to control and test, since excessive air can severely reduce strength and pumpability. Entrained air is generally considered undesirable in mixes of high cement (or other fines) content where frost resistance is not required. However, the authors have used entrained air to provide lubrication in mixes where fines were excessive and strength relatively unimportant.

Relationship between the durability factor and hardened air content of mixes with Vinsol resin admixture (Set 1) or synthetic admixture (Set 2) is shown in Figure 4.1. This demonstrates that examination of the air void properties is a good idea for new products, which in fact may only be appropriate as a car washing detergent.

Many investigations show that entrained air is still necessary for resistance to freezing and thawing, even in very high-strength concrete. The authors are dubious about this, considering that it may only apply to fully saturated specimens used in laboratory investigations with the extreme freeze–thaw cycling rather than to real structures. There are plenty of examples of structures that have performed well in freeze–thaw environments without air entrainment. However, the omission of entrained air in concrete subject to freezing and thawing represents a risk.

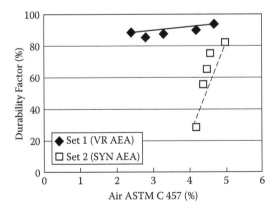

Figure 4.1 Relationship between the durability factor and hardened air content of mixes with Vinsol resin admixture (Set 1) or synthetic admixture (Set 2). (From U.S. Federal Highway Administration, 2006.)

4.3.6 Water-resisting admixtures

One important limitation of conventional concrete, even of good quality, is the presence of microcracks, capillaries, and microcapillaries into which water is able to penetrate, sucked in by surface tension or driven by an external hydrostatic pressure. Where concrete is in contact with damp soil or is below the water table, water is drawn through the concrete into the structure. This can lead to unacceptable dampness as well as damage to carpets, furnishing, and equipment. Water transmission under such conditions can also result in the dangerous accumulation of aggressive salts in the concrete leading to corrosion of the reinforcing steel and deterioration of the concrete itself. Similar problems occur in concrete exposed to periodic wetting with water containing salt or other aggressive agents such as in the splash zone or bridge decks. Therefore, control of water (salt) movement is often vital to achieve the required performance and durability.

In the past designers attempted to isolate the concrete from water by the use of membranes or surface coatings. However, it is extremely difficult, if not impossible, to ensure there are no weak points or faults through which water along with any dissolved salts or acids can penetrate leading to leakage, dampness, and possibly corrosion. As a result of these problems, attention is now focused on water-resisting admixtures to control water and moisture movement as well as improving concrete durability.

Unfortunately, the information available on this class of admixture has been full of generalisations based on little, if any, controlled data. A Building Research Advisory Board report (1958) said that in the opinion of the majority of 61 observers, "dampproofing admixtures are not ... effective or acceptable in controlling moisture migration through slabs-on-ground". On the other hand, Robery (1987), based on work on a hydrophobic pore-blocking ingredient (HPI), takes a more positive view: "Hydrophobic additives will produce the most marked improvement in average quality concrete subjected to low hydrostatic pressures (<10 m head). The waterproofed concrete thus competes directly with conventional tanking and roof membrane systems". Rixom and Mailvaganam (1986) in their summary of concrete hydrophobic admixtures also support this more positive view. Comparative testing has shown significant variation in short- and long-term performance of hydrophobic admixtures, which has led to the conflicting views on this class of admixtures in the literature.

ACI 212.3R-10 (Table 4.1) refers to permeability reducing admixtures (PRAs) and subdivides them into nonhydrostatic (PRAN) and hydrostatic (PRAH). As permeability is defined as water flow due to a hydraulic gradient, water permeability necessarily involves the hydraulic pressure. Accordingly, the attempt to create a subcategory of permeability reducing admixtures for nonhydrostatic conditions is technically invalid.

The classification used in BS EN 934-2 is water resisting admixture, which would be a technically correct to refer to water resisting admixtures for hydrostatic and nonhydrostatic conditions. Water resisting admixtures are divided into crystalline, hydrophobic, and hydrophobic pore blocking.

Crystalline admixtures were developed from surface applied products that were designed to penetrate voids and cracks in the concrete. The suppliers claim that the crystals, which accelerate the autogenous healing capabilities of concrete, are able to grow to fill and block static cracks up to 0.4 mm. The silicate reacts with calcium hydroxide (produced by the cement hydration process) to form a calcium silicate hydrate (C-S-H) similar to that formed by cement hydration but with a variable hydrate concentration (CSHn) (Trinder, 2000). Although much of the information available on these products is proprietary, scanning electron micrographs show crystal formation in capillary pores suggesting that they should be able to fill fine cracks and voids provided the conditions are appropriate. Mitsuki et al. (1992) found crystal growth occurring and concluded from qualitative analysis of the concrete by an energy scattering x-ray analysis procedure that the needlelike crystals were C-S-H. Anecdotal evidence suggests beneficial effects of crystalline admixtures should be at least as good as traditional autogenous healing.

There are a number of studies that indicate improved chemical resistance with the use of such products. For example, Trinder et al. (1999) found that the addition of a crystalline admixture substantially improved resistance to ammonium sulfate compared to a reference concrete. Yodmalai et al. (2009) showed reduced surface chloride content and apparent chloride diffusion coefficient for both the crystalline admixture and surface treatment. These short-term studies suggest that the resultant crystals appear to be durable and of some benefit to the concrete. Although the longer-term durability of such crystals has not been independently established, C-S-H crystals (the basic chemistry of the crystals and cementitious hydration) do have a long history. The problem of evidence for longer-term durability exists with many materials. Tests on penetrability properties of concrete (absorption, permeability, and chloride diffusion) have tended to show limited effect of crystalline admixtures on these parameters compared to a comparable reference concrete of reasonable quality.

In the opinion of the authors, the primary advantage of crystalline admixtures is in enhancing autogenous healing of cracks and voids in concrete, which is an important component in achieving watertightness in real structure. Although there appears to be little evidence of reduced penetrability in higher quality concrete, there is evidence of improved chemical resistance.

The next type of water resisting admixture is the hydrophobic admixture. Hydrophobic admixtures are believed to form a thin water-repellent layer within the pores and voids in the concrete matrix that exhibit high contact angles (θ) to water. Rixom and Mailvaganam (1986) suggest that concrete containing hydrophobic admixtures can be considered to have a contact

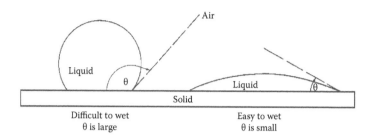

Figure 4.2 Effect of contact angle on the "wettability" of a surface. (From Rixom, M. R., and Mailvaganan, N.P., *Chemical Admixtures for Concrete*, E & F Spon, London, UK, 1986.)

angle of approximately 120° based on that surfaces coated with waxes or fatty acids would have this contact angle (Figure 4.2).

Using a contact angle of 120° and maximum capillary size of 500 nm, Rixom and Mailvaganan (1986) calculate the theoretical pressure necessary to penetrate the concrete would be 14 metres. In fact, wetting still does occur due to defects in the hydrophobic film or larger voids within the matrix. Research by Yiannos (1961) showed a monolayer of stearic acid on copper had a contact angle of 104° and this would be a more conservative estimate. The hydrophobic pore-blocking admixtures combine a hydrophobic material with a polymeric material, which coalesces under pressure to form a plug.

The conflicting claims regarding different proprietary products makes it difficult for specifiers and users to assess this group of admixtures. Figure 15.5 in ACI 212.3R highlights the problem. The coefficients of permeability for the reference samples as reported in the three BBA certificates are 2.2×10^{-12} m/s, 2.0×10^{-13} m/s, and 4.3×10^{-14} m/s indicating that the concretes were of markedly different quality. Accordingly the relative values given in the figure have little real meaning. The second permeability result given in the British Board of Agreement (BBA) certificate for the hydrophobic pore-blocking admixture, which shows an order of magnitude reduction, is not shown in the figure or table. Clearly, data on the relative permeability of the same concrete mixture containing different admixtures would be more helpful.

The comprehensive test program conducted by Roberts and Adderson of the Building Research Establishment (BRE) in 1985 provides dramatic evidence of the vast variability in short-term performance of concrete containing different hydrophobic admixtures. Five out of the nine hydrophobic admixtures tested (and the water reducing admixture) actually increased water penetration compared with the control in 10 or more of the 15 combinations of test method and curing procedure. Only one admixture (a hydrophobic pore-blocking ingredient) consistently reduced water penetration under all five test methods. Although two admixtures gave significant reductions compared with the control in the absorption tests (namely, ISAT, and capillarity and water immersion), they did not consistently reduce water penetration in the pressure tests (namely, DIN 1048 and vacuum soaking) (Aldred, 1989). These would be considered nonhydrostatic water resisting admixtures. Roy et al. (1995) also found significant variability in short-term penetrability of the different water resisting admixtures tested. Again the same hydrophobic pore-blocking ingredient was the only product that consistently reduced penetrability.

Another important aspect of water resisting admixtures is long-term performance. Neville (1981) mentions that "water-proofing admixtures ... are supposed to repel water by an electrostatic charge which they form after reacting with calcium ions on the walls of the capillaries in the hydrated cement paste. It is doubtful whether this effect persists over long periods". This is a valid concern and there are examples of hydrophobic admixtures that have not provided acceptable long-term performance. Only water resisting admixtures where the chemistry has been proven to achieve long-term performance should be considered as a suitable "waterproofing" or durability enhancing systems.

In addition to reducing penetrability, effective admixtures appear to assist in reducing cracking by reducing interfacial surface tension (similar to shrinkage reducing admixtures). Another benefit in real structures appears to be limiting lateral water movement from cracks and voids, which makes localised repair easier.

The use of fly ash, ground-granulated blast-furnace slag (GGBS), silica fume, and other supplementary cementitious material (SCMs) reduces permeability. These materials have been dealt with in Chapter 2. It is interesting that in spite of the widespread use of SCMs that significantly reduce permeability, there is a growing trend toward using water resisting admixtures. This appears to be due to ease with which faults in the concrete can be repaired and the preparedness of better suppliers to warranty performance. It is also probably due to continued problems with membrane waterproofing.

4.3.7 Shrinkage reducing admixtures

Capillary tension theory is the leading theory to explain autogenous and drying shrinkage of concrete. Shrinkage reducing admixtures (SRAs) are typically polyoxyalkylene alkyl ether or similar (ACI 212.3R) which reduce surface tension thereby reducing the tension that develops in the capillaries during drying (either to the environment or due to hydration). Bentz et al. (2001) measured the effect of the addition of a 6% solution of a dipropylene glycol ether blend in water. The surface tension was reduced by 57% compared to distilled water.

The reduction in autogenous and drying shrinkage is reportedly up to 50% or more using these admixtures. Therefore this type of admixture can be an important tool for designers and contractors to be able to increase the size of concrete panels and to reduce or eliminate cracking in heavily restrained concrete. Reducing autogenous shrinkage using SRA is very helpful in higher strength concrete with low w/cm. The problem with autogenous shrinkage is that it is a through section strain and usually compounded by thermal shrinkage.

By changing surface tension, SRAs change other properties of the concrete. Bentz et al. (2001) found that the desorptivity was decreased and resultant drying profile changed from uniform drying to a sharp front. Aldred (2008) found that SRA reduced sorptivity, desorptivity, and wick action.

Shrinkage can also be reduced by shrinkage compensators such as finely divided iron or calcium sulphoaluminate. These materials work but require careful use to avoid the expansion tendency being disruptive. Also it must be remembered that they do not actually work by reducing shrinkage. In both cases an expansion is produced while the concrete is kept damp (i.e., before any shrinkage occurs) and the concrete then shrinks normally. The initial expansive tendency is restrained by reinforcement or by abutting concrete and develops a compression that dies away under the later influence of shrinkage. In addition to the risk of excessive expansion causing disruption, there can also be a threshold effect in which the expansive tendency is inadequate and the precompression is all lost in creep of the concrete, leaving no effect on subsequent shrinkage.

In the United States shrinkage compensating cements are available and even expanding cements designed to automatically apply prestress to cast-in steel tendons. This is done by the incorporation of calcium sulphoaluminate in the cement during manufacture.

4.3.8 Viscosity modifying agents (VMAs)

Viscosity modifying agents (VMAs) include

- Wax emulsions
- Thickening agents (methyl cellulose, polyethylene oxide)
- Fly ash
- Silica fume

Wax emulsions and thickening agents do improve pumpability, but the improvement is not dramatic and the expense and difficulty may be appreciable. Fly ash is very useful if available. Silica fume at low replacement levels of 2% to 3% is very effective.

It has been said that the only satisfactory test for pumpability is to pump the concrete. However a most effective cheap and simple test is bleeding. It is probably true that concrete that bleeds excessively will not pump, but the reverse is not necessarily the case. Too high a workability can be as harmful as too low a workability if the mix has inadequate cohesion. It can be seen that the aforementioned admixtures are all in effect bleeding suppressants. This old dictum has recently been taken to a new level by Kaplan, de Larrard, and Sedran (2005) in a research project involving a specially assembled 148 m closed circuit of piping and over 60 truckloads of concrete. A technique using a standard pressure air metre with tetra-chloroethylene instead of water and so measure water squeezed from the concrete was employed to measure bleeding under pressure. It was found that the rate rather than the quantity of bleeding was significant. This was fortunate because it allowed a rapid result to be obtained. The pumping procedure was also found to be very important. Avoiding delays between trucks, defective joints in the line, and pumping slowly during priming as well as when difficulties are experienced is well-known advice. An interesting new observation was the importance of the first concrete intermingling with the priming slurry if it was too fluid. This confirms our collective experience that a cohesive slurry, preferably similar to the concrete without aggregate is required.

VMAs are particularly important when SCC of relatively low strength (and therefore low cementitious content) is involved. SCC has been reported to demonstrate excellent pumpability on the Eureka building in Melbourne, Australia, currently the tallest in the Southern hemisphere.

4.3.9 Self-consolidating concrete (SCC)

Self-consolidating concrete (SCC) was discussed as a current hot topic in the third edition. It was highlighted that its ranking in terms of production volume was nothing like its ranking in volume of technical literature. However, there has been a considerable increase in the volume of SCC used, particularly in the precast industry where the advantages of ease of placement and high-quality finish have made SCC the preferred concrete used in many markets. The greater quality control required is easier to achieve under the factory conditions of a precast plant than in a standard premix plant. There has also been greater resistance to the use of SCC for in situ construction because of the increased materials cost as a result of the higher binder content and admixture dosage. One region where this has not been a significant issue has been in foundation elements in the Middle East where the durability requirements of the specification would normally require a low water to cementitious ratio and a high binder content, often with significant replacement with supplementary cementitious materials. Therefore modification of aggregate grading and possibly the use of suitable viscosity modifying agent may be the only changes to achieve self-consolidating characteristics at minimal additional cost. The fact that SCC could greatly reduce the placement time and the size of the concrete placing team means that SCC could provide significant cost savings in these applications.

The improved capability of superplasticisers has played an important role in expanding the capability of SCC to achieve the required fresh and hardened properties. High fine aggregate percentage and often reduced maximum aggregate size coupled with low w/cm and high workability would only be possible with considerable dosage of an effective superplasticiser. The advances in viscosity modifying or viscosity enhancing admixtures (VMAs or VEAs) are potentially more important as they help to ensure adequate segregation resistance, even with variation in aggregate grading. The active marketing of VMAs to provide robust SCC with lower binder contents is expected to further expand the application of SCC for in situ applications. One of the main points for advocating SCC has been to reduce the cost of repairing honeycombing and other defects in concrete elements. As the admixture suppliers often sell the repair products, it is commendable that they are trying to eliminate these defects (but then again the repair products are probably in another division!). Olafur Wallevik has done a great deal of research on low binder content SCC and has been able to achieve 40 MPa SCC with only 280 kg/m^3 binder.

Improved quality control procedures, greater acceptance of simple practical test procedures to assess segregation resistance (see Chapter 7), and the development of standardised SCC mixes by premix suppliers will also help expand the application of SCC for in situ construction. Educating specifiers in the advantages and limitations of SCC is also an important factor.

As most of the test procedures on SCC focus on passing or filling ability, there can be a tendency to add too much superplasticiser to reduce the viscosity. This can lead to reduced segregation resistance, particularly if the retardation level is high. We would suggest that segregation resistance is at least as important as adequate flow and is rarely assessed on site. For example, on one project a concrete mix was modified to reduce the VMA dosage and increase the superplasticiser content, as the L-box value was less than the guideline given by EFNARC. A site trial was conducted, which showed no segregation. However segregation did occur during initial casting. Further trial mixes demonstrated a tendency for the slump flow of the mix to increase over time and the 5 minute V funnel increased markedly after 90 minutes indicating reduced segregation resistance. This suggested a limited effective duration for the VMA, which was not compensated by hydration effects due to the increased retardation. This was an example of the dangers of using arbitrary performance criteria to define the concrete mix design without a full appreciation of the interaction of the different parameters. Testing procedures for SCC together with their advantages and limitations are discussed in Chapter 7.

An interesting historical note is that Thomas Edison was very interested in concrete and set up a concrete company. He was convinced that self-consolidating concrete was the way forward (Figure 4.3). He envisioned whole structures being pumped from one point. He did not have the tools of superplasticisers, and VMAs but he had the vision and would be pleased to see the progress being made.

Figure 4.3 One of Edison's concrete houses under construction in 1919.

4.3.10 Corrosion inhibitors

As most durability problems in concrete are related to chloride-induced corrosion of embedded reinforcement, it is not surprising that admixtures would be developed to increase the chloride threshold level. Although there are many chemicals that have corrosion inhibiting properties, there are basically four generic types commercially available: calcium nitrite, amine carboxylate, amine-ester organic emulsion, and organic alkenyl dicarboxylic acid salt.

Apparently sodium nitrite was used by the Germans to reduce the corrosive effects of calcium chloride, which was used as an accelerator when constructing the infrastructure to mount the Blitzkrieg at the start of WWII. However, sodium nitrite tends to reduce strength. In the 1980s, the company WR Grace began marketing calcium nitrite as an anodic inhibitor and it is the most widely used material today. Calcium nitrite is also an effective accelerating admixture.

There are examples of long-term corrosion resistance using calcium nitrite. Some concerns exist over the increased cracking potential of concrete containing large quantities of calcium nitrite due to the increased shrinkage and temperature rise. Montes et al. (2004) showed that the effect of calcium nitrite on corrosion inhibition in cracked elements was limited, presumably due to its lower resistivity. Ann et al. (2006) showed that this lower resistivity tended to offset some of the benefit of the inhibitor.

Amine carboxylate admixtures were developed from vapour phase inhibitors, which have a long history of use in other industries. The initial applications for reinforced concrete were based on surface treatment and the expectation of migration to the reinforcement. This always seemed a bit doubtful in better quality concrete with reasonable cover. Molecules would be expected to take the easy option of dissipating into the air rather than the tortuous journey through the dense concrete matrix. However, they do bond tenaciously to reinforcing steel. The supplier can test whether surface application will be effective. As an admixture, however, it does seem to have a number of advantages. It is both an anodic and cathodic inhibitor, which is useful in both new construction and repair applications. It does not detrimentally affect the fresh or hardened properties of the concrete, but there is some retardation. Recent data from a bridge deck in Minnesota that was poured in 1986 demonstrates a significant reduction in corrosion rate in a good quality concrete.

Both amine-ester organic emulsion and organic alkenyl dicarboxylic acid salt act as water resisting admixtures and therefore it is difficult to assess their claims to be corrosion inhibitors as the test procedures to demonstrate their performance normally involve penetration of waterborne chlorides followed by corrosion initiation and propagation.

4.3.11 Workability retaining

Workability retaining admixtures are added separately to concrete that may already contain a water reducing or HRWR admixture. These admixtures extend the time that concrete can be workable without affecting subsequent setting time or early compressive strength development. The polymers within these admixtures are slowly released coating the surfaces of the cement particles preventing agglomeration and stiffening. They are a useful tool to tailor concrete for particular applications without going to the extent of using hydration controlling admixtures, which would be better suited for prolonged delay in setting and reactivation.

Chapter 5

Properties of concrete

Before starting to design (or specify) concrete, it is necessary to consider what properties we want the concrete to have and also what properties we do not want it to have. Some properties may come under both headings, such as heat generation, but generally undesirable properties are simply a lack of desirable properties.

Important properties include

- Durability
- Strength
- Water/ion transport
- Rheology
- Dimensional stability
- Good appearance
- Economy
- Sustainability

5.1 DURABILITY

Durability must come first on our list because if our concrete does not achieve the required design life, it cannot display any of the other desirable properties (not even economy because the most expensive concrete you can get is that which has to be replaced!). However there is a difference between durability for a few years, a few decades, or a few centuries, between durability at any price and "reasonable" durability of economical concrete, and durability in benign or aggressive environments. More particularly there is a difference between the durability of plain concrete and the durability of reinforced concrete.

5.1.1 Corrosion of reinforcement

Generally, reinforcement is the Achilles' heel of concrete. We are all familiar with cracked, rust-stained concrete caused by the expansion of reinforcing steel. Roman concrete was not reinforced and this is a major reason for its survival for centuries. However, as we shall see, there can be durability problems with unreinforced concrete aside from structural issues.

The major factor in the corrosion of reinforcing steel is the thickness and quality of the concrete cover. In theory, without adequate cover, concrete cannot protect the reinforcement. However excessive cover means that the surface concrete is essentially unreinforced and can crack due to thermal stresses or shrinkage, sometimes with the reinforcing cage acting as a crack inducer. Therefore, specifications that call for 100 mm cover or more may result in no effective cover whatsoever! The good durability of spun pipes, ferrocement, and steel fibre reinforced concrete with limited or negligible cover to the reinforcement highlights that corrosion is a complicated subject and there can be exceptions to the general rules.

With reasonable cover, the next factor is the penetrability of the concrete. "Penetrability" is a collective term describing the transport properties of concrete, including absorption, permeability, and diffusion. Basically steel will not rust unless water and oxygen can reach it and it has been depassivated by chloride ions or carbonation. Since it is the alkalinity of cement that provides passivation of the steel, and since a lower water to cement ratio (w/c) tends to reduce penetrability, it used to be thought that a high Portland cement content was the appropriate way to achieve durability. The substitution of a proportion of fly ash or blast-furnace slag for some of the cement was considered to reduce durability. It is now realised that substitution of cementitious materials generally reduces penetrability and is an important positive factor in reducing corrosion. However, good curing can be even more important with concrete containing fly ash or slag to be effective. There are also a wide range of admixtures and technologies that can be used to reduce corrosion of reinforced concrete.

An exception to the general rule regarding corrosion can be carbonation. Carbonation occurs when penetrating carbon dioxide dissolves in the pore water and reacts with calcium hydroxide reducing the alkalinity of the matrix. The phenomenon is sensitive to the internal relative humidity of the concrete. The conditions that lead to higher rates of carbonation where the internal relative humidity is between 50% and 70% would not be expected to cause significant corrosion in the event of the carbonation front reaching the reinforcing steel due to the lack of moisture and higher resistivity. The propagation phase of the reinforcement corrosion due to carbonation could be 100 years. Accordingly a 100-year design life could be assured even if the concrete was carbonated before removing the formwork! In situations where the concrete can carbonate and then have access to moisture can have more rapid rates of corrosion. Vehicular tunnels with high carbon dioxide concentration at the internal surface and moisture availability to the reinforcing due to wick action may require extra care as well as structures exposed to periodic wetting. Carbonation certainly makes potential chloride-induced corrosion worse by releasing bound chloride as well as reducing alkalinity.

Because of the sensitivity to the moisture content, the possible difference in carbonation rates between a lab test and site exposure can be very large. Where practical if there is a dispute over the potential carbonation, the authors would suggest periodic sampling to measure in situ carbonation over the coming few years. If in situ rates show that there is potential for carbonation through cover within the proposed design life, the concrete can be coated later. Carbonation is a slow process and therefore the time to intervene to prevent the carbonation front reaching the reinforcement is long.

5.1.2 Alkali–aggregate reaction

An important cause of deterioration in concrete is alkali–aggregate reaction, which can include both alkali–carbonate reaction (ACR) and alkali–silica reaction (ASR). Reactive carbonate rocks are relatively rare; if they are suspected, potential ACR should be established by testing to ASTM C586, and if present, they should be avoided. Alkali–silica reaction is a disruptive expansion of the cement matrix arising from the combination of alkalies (usually, but not necessarily solely, from the cement) and reactive silica within the aggregate. Although generally confined to particular geographies, the phenomenon can be disastrous when it does occur. There are three possible strategies to limit its occurrence. One is to limit the quantity of total alkalies (sodium and potassium) in the cement to less than 0.6% calculated as Na_2O equivalent ($1 \times Na_2O + 0.685\ K_2O$). Another is to test the aggregate for reactivity. A third possibility is to provide an excess of reactive silica in the form of fly ash, silica fume, or natural pozzolan so as to consume excess alkali present in a nonexpansive surface reaction product. Iceland, which had widespread problems with ASR due to reactive aggregates and a high alkali cement, has virtually eliminated ASR since 1979 with the introduction of 7.5% silica fume into its cement as well as washing of sea dredged materials and limiting the use of reactive materials. Queensland in Australia has also largely solved its ASR issues by requiring a minimum fly ash replacement of 20% of the component Portland cement.

5.1.3 Sulfate attack

Sulfate attack is an important deterioration mechanism of the concrete itself. Sulfates react with the portlandite and tricalcium aluminate in hydrated concrete to cause disruptive expansion. Sulfate resisting (Type V) cement has a limited tricalcium aluminate content. Low heat Portland

(Type IV) cement also has its tricalcium aluminate content limited for the different reason that it generates more heat. However, both of these cements have lower than normal resistance to penetration by chlorides, so neither should be used in marine situations because seawater contains both sulfates and chlorides. A better solution is to use fly ash or blast-furnace slag replacement. The latter is particularly suitable for marine use especially at higher replacement levels. Fly ash and ground-granulated blast-furnace slag (GGBS) replacement not only increase resistance to sulfate and chloride ions but also can significantly reduce temperature rise. Silica fume is also very effective in reducing penetrability. Surface carbonation has been found to improve sulfate resistance. In the presence of magnesium sulfate, the cement paste itself is further weakened as the calcium silicate hydrate is decalcified. This form of attack is more aggressive and requires greater precautions. BRE Special Digest 1 gives good advice.

5.1.4 Delayed ettringite formation

Although most durability issues are associated with corrosion of reinforcement, delayed ettringite formation (DEF) has resulted in 10 mm cracks in the unreinforced blocks of a port facility within 4 years of service. There is debate about the exact mechanism of DEF. The generally accepted view is Portland cement concrete, which attains a temperature of approximately 70°C or more during hydration, appears to inhibit the formation of preliminary nonexpansive ettringite. The unreacted sulfate within the concrete is then available to react with tricalcium aluminate in the cement similar to traditional sulfate attack. Most examples of disruptive DEF have occurred in moist environments such as ports, dams, and railway sleepers.

There are many factors that have been found to influence the susceptibility to DEF: sulfate to aluminate ratio, cementitious replacements, cement fineness, and reactivity. As mentioned in Chapter 6, the authors would advocate a peak temperature limit of 70°C regardless of other considerations is probably the best way to help ensure that DEF is not a problem. DEF may not be common, but it can cause severe problems that can be easily avoided.

5.1.5 Thaumasite

In concrete exposed to sulfate, carbonate, and water at low temperature (less than approximately 15°C) the calcium silicate hydrate crystals can convert to thaumasite. Accordingly, the binder within the concrete can be turned into a weak friable material. The source of carbonate may be calcareous aggregates, superfine $CaCO_3$, or dissolved carbonate in the water. One of the greatest threats of thaumasite attack is friction piles where a surface thaumasite attack would effectively eliminate the friction between the pile and the ground. BRE Special Digest 1 provides a full discussion of the problem and recommendations. GGBS is considered particularly helpful for thaumasite resistance (Neville, 2011). Surface carbonation has been found to improve resistance to thaumasite.

5.1.6 Physical salt attack

Physical salt attack is the deterioration of concrete (and masonry) due to the accumulation of salt due to evaporation. This can be a problem, particularly where there are high concentrations of sodium sulfate. The problem can also be called salt weathering, salt damp, salt crystallisation or physical salt distress. Physical salt attack is considered to be caused by the cycling of sodium sulfate between its anhydrous and hydrated forms, which results in an expansion of over three times. Other salts may also be involved. Other researchers have suggested that it is caused by the supersaturated salt solutions. The key parameter involved in the accumulation of the salts appears to be sorptivity.

5.1.7 Chemical attack

Portland cement based concrete is not resistant to acid, although concrete with a low penetrability will be attacked less rapidly. Supplementary cementitious materials tend to improve chemical resistance. Supersulfated cement, which contains 80% to 95% GGBS, has been found to be resistant to acidic conditions down to a pH of 3.5. Unfortunately supersulfated cement is no longer readily available. Surface carbonation has been found to improve acid resistance.

High alumina or calcium aluminate cements have been found to be resistant to acids and a range of other chemicals. Provided they are used in accordance with the manufacture's recommendations to limit conversion issues, they are a suitable material and are making a resurgence in sewer linings. Interestingly, some geopolymer concrete is also resistant to acid attack.

Proven high performance water resisting admixtures have been found to greatly increase resistance of Portland cement concrete to a range of chemicals.

Distilled or soft water will also attack Portland cement based concrete by leaching the calcium hydroxide and then decalcification of the calcium silicate hydrate (C-S-H). Fertilisers such as ammonium nitrate also tend to leach Portland cement concrete and cause progressive deterioration.

5.1.8 Freeze–thaw attack

Under conditions of freezing and thawing, concrete can suffer significant damage, particularly if the water contains salt. The typical solution to potential freeze–thaw damage is air entrainment, which has been effective over the years. One bone of contention is whether high strength concrete still requires air entrainment for frost resistance due to the small pore sizes rendering the water not freezable. There is no question that test cylinders cured in a water tank and frozen while saturated will show a benefit from air entrainment in even very high strength concrete. However the self-desiccation and the difficulty of resaturation would seem to suggest that air entrainment may not be necessary.

5.2 MECHANICAL PROPERTIES

5.2.1 Compressive strength

Compressive strength is well established as the primary criterion of concrete quality. Mix design has generally meant designing a mix to provide a given strength. Although strength is often not the most important requirement, the reason for its use as a performance criterion is clearly shown by the step following its selection in most mix design procedures. This is to convert the strength requirement into a water to cement ratio. The relationship between strength and w/c is generally attributed to Abrams (1929). Actually Féret (1896) preceded him and proposed a more accurate proportionality, that between strength and the ratio of cement to water plus voids (Neville, 2011). It may be that accuracy was not the important thing, partly because the w/c itself was arguably more important than the strength it was assumed to represent. Partly because the simplicity of the concept was as important as its accuracy.

Although the concept of w/c is simple, and its approximate implementation is also simple, it would be a difficult criterion to enforce by testing. An accurate way of establishing the w/c ratio of a given sample of production concrete (of which the w/c ratio versus strength relationship has already been established) is to test its strength. It is perhaps unfortunate that w/c ratio rather than c/w ratio came to be the popular parameter since, over a substantial range, strength has an almost linear relationship with c/w ratio. So much of the importance of strength is as a test method and a means of specification for w/c ratio.

A primitive way of designing a mix, assuming that only one fine and one coarse aggregate were involved, would be to make a mix of any reasonable proportions (say 1:2:4) and fairly high slump (say, 100 mm). If a sample of this concrete were heavily vibrated for several (say, 15) minutes in a sturdy container (such as a bucket, not as small as a cylinder mold) then any excess of either coarse aggregate or mortar would be left on top. If the top half were discarded, then the proportions of the bottom half would be a reasonable guide to the desirable sand percentage to use. This is a useful exercise for students since it illustrates the concept of filling the voids in the coarse aggregate with mortar and demonstrates that an ideal mix cannot be overvibrated once it is fully compacted in place (in that the remaining concrete will not further segregate however long it is vibrated).

Very high strength depends on a number of other things besides w/c ratio. These include the strength of the coarse aggregate, and the bond between the matrix and the coarse aggregate. It used to be very difficult to achieve a strength much in excess of 90 MPa (13,000 psi). Strengths of double this amount can be obtained given a strong coarse aggregate, silica fume, and a superplasticising admixture. Day recalls carrying out trial mixes for 60 MPa concrete in the late 1970s before either silica fume or superplasticiser were available. Of the two coarse aggregates tried, the stronger one gave unsatisfactory results. This was because it was such a hard impermeable material that the matrix did not bond to it sufficiently. With silica fume and superplasticising admixtures now available, excellent bond was developed and the stronger coarse aggregate gives better results than the other and both can easily exceed 100 MPa.

There are two words of caution about using very high concrete strengths. One is that concrete in a structure cannot be saturated with water as can test cylinders or cubes in a water bath. It will have a w/c insufficient to provide full hydration and will therefore self-desiccate and not develop the full strength of the test specimens. At best it may be possible to prevent the loss of any of the mixing water by polythene wrapping immediately on demolding or placing the concrete in permanent formwork such as a steel pipe column. So perhaps high strength test specimens should be polythene wrapped rather than water-bath cured, although this should probably be restricted to a few comparison tests, since it may be undesirable for quality control from the viewpoint of introducing variability into the results. The opposite problem occurs when the high strength test specimens dry out due to poor sampling and early protection. Because of the low penetrability, the specimens do not absorb water on immersion. The test specimens may give satisfactory early strength but significantly reduced strength compared to the in situ concrete at later stages. The provision of saturated lightweight particles in a mix to provide internally the water for curing (Bentz et al., 2005) helps maximise the performance of very high strength concrete and also helps address the problem of autogenous shrinkage. Another suggestion has been to use a proportion of reactive magnesia to perform a similar function (see Chapter 13).

The other problem with very high strength concrete (actually very low penetrability concrete) is that of explosive failure in a fire situation.

The theory is that water vapour from the interior will be unable to escape and will cause explosive spalling. This may seem unlikely considering the self-desiccation referred to earlier, but in fact chemically combined water can be driven off. Nylon or polypylene fibres introduced to the mix melt and provide an escape path for moisture. Generally, structures fail in a fire more due to a failure to protect the steel than from deterioration of the concrete, so lightweight aggregate concrete, providing better thermal insulation, will show an improved result.

5.2.2 Tensile strength

Concrete is relatively weak in tension, which is the reason for the use of reinforcement in most concrete. Cracking in concrete will occur when the stress exceeds the tensile strength and therefore this property is specified in different applications. Indirect tensile tests or "Brazil splitting test" is the most common procedure and can give quite consistent results in a good laboratory. Direct tensile tests, on the other hand, are difficult to conduct without causing eccentric stresses, which results in high variability and unrealistically low values. As with many other test procedures, the frequency of tensile strength testing is generally significantly less than compressive strength testing and may not be conducted by the premix company for in-house quality control. Therefore production quality control should be based on an established relationship between the compressive strength and tensile strength using the proposed materials.

5.2.3 Flexural strength

Flexural strength or modulus of rupture is an important property, particularly in pavements where it is often specified. Flexural strength is generally measured by three-point loading of a beam. Flexural strength can be more affected by changes in aggregate properties than compressive strength. A good laboratory can achieve low variability results that can be used directly for quality control but often its relationship with compressive strength would be used for quality control by the premix company.

5.2.4 Modulus of elasticity

Modulus of elasticity gives an indication of the stress–strain behaviour of concrete. The Young's modulus of elasticity is calculated from the linear part of the stress–strain curve or the initial tangent (Neville, 2011). There are some situations where it is good for the concrete to have a low modulus such as roads or dams where strains caused by settlement can be accommodated without cracking. Many super tall structures are designed based on stiffness and therefore a high elastic modulus may be specified. Elastic modulus is strongly influenced by the modulus of the aggregate. The relationship between compressive strength and modulus will vary based on the aggregate type used. The Burj Khalifa required an elastic modulus of 43.8 GPa at 90 days. The average value achieved was 49.5 GPa with a standard deviation of 2.5 GPa for a concrete with an average compressive strength of approximately 110 MPa. Other projects using lower modulus aggregates may require higher compressive strength to achieve the required modulus.

Elastic modulus is more difficult to measure than compressive strength and will generally be infrequently tested. Therefore production quality control should be based on an established relationship between the compressive strength for the specific materials used and the measured Young's modulus. One area of contention is whether elastic modulus should be specified as a characteristic value or an average. As stiffness is generally the property of an element as a whole, some believe that the average is the suitable design requirement.

5.3 TRANSPORT PROPERTIES

Water penetration into concrete and resultant accumulation of dissolved salts are important problems facing the concrete industry. The capillary pore system within the hydrated cement paste, the coarse porosity of the aggregate–matrix interface together with any microcracks provide pathways for the transport of water and any dissolved chloride ions. RILEM (1995) outline three distinct mechanisms of water transport:

- Permeation through concrete under a hydraulic gradient
- Sorption into the unsaturated concrete
- Diffusion of water vapour and ions under a concentration gradient

These may act singly, simultaneously, or in series depending on the exposure condition and the moisture content of the concrete. Under most exposure conditions, water and dissolved ions penetrate concrete by more than one transport mechanisms. Laboratory examination of these mechanisms often involves controlling the experimental conditions such that a specific transport parameter is isolated from others enabling the relevant transport coefficient to be calculated from existing theoretical models (RILEM, 1995). Once the individual transport parameters have been established, some models have been developed in an attempt to predict the actual water and chloride flow resulting from their interaction, for example, Stadium.

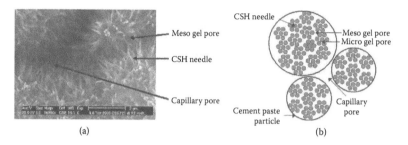

(a)   (b)

Figure 5.1 (a) C$_3$S after 21 days curing time, CSH-needles with diameter approximately 5 nm. (From Espinosa, R. M., and Franke, L., *Cement Concrete Res.*, 36, 1956–1970, 2006b.) (b) Scheme of the pore structure of hardened cement paste. (From Jennings, H.M., *Cement Concrete Res.*, 36, 101–116, 2000.)

5.3.1 Porosity

Porosity is an important characteristic of concrete that influences many aspects of its behaviour: mechanical properties, transport properties, and durability. For cementitious materials, Espinosa and Franke (2006a) define pores with hydraulic radii as follows (Figure 5.1):

- Micro-gel pores, <1.0 nm
- Meso-gel pores, 1.0–25 nm
- Micro-capillary pores/meso-capillary pores, 25–50 nm
- Macro-capillary pores, 50 nm–1 µm

The simplest measure of porosity is the total voids content as estimated by the ASTM C642 Volume of Permeable Voids test. More sophisticated techniques measure pore size distribution. Mercury intrusion porosimetry (MIP) has an advantage over other testing techniques, such as a capillary condensation in being able to measure pore radii ranging from a few nanometers to several hundred micrometers (Diamond, 1971). Mercury is a nonwetting liquid for cementitious materials and consequently it has to be forced into the pores. Pore size and volume quantification are calculated from the pressure required to force the mercury into the sample. Although a valuable tool, Diamond (2000) suggested that it was inappropriate method for the absolute measurement of pore size distributions in "cement-based materials primarily because of the so-called" ink-bottle effect. This refers to a larger void, which is accessed through a smaller pore.

5.3.2 Permeability

The flow of water (with or without chloride ions) caused by a pressure head is water permeability. In this transport mechanism flow through the capillary system is assumed to be laminar (i.e., a steady-state condition has been established). The coefficient of water permeability is calculated according to D'Arcy's equation. Considering the tortuous nature of the capillary network within concrete, particularly of high quality, the time to achieve a steady-state condition may be very long. Accordingly, a coefficient of water permeability is sometimes calculated from uniaxial penetration based on Valenta's equation. There are three avenues by which water can penetrate through concrete under pressure:

1. Gross voids arising from incomplete compaction or segregation
2. Micro (or macro) cracks resulting from plastic, autogenous, or drying shrinkage; thermal stresses; or plastic settlement
3. Pores or capillaries resulting from mixing water in excess of approximately that which can combine with the cement, that is, water in excess of 0.38 by mass of cement

Gross voids may be regarded as too obvious a cause to be included. However, they are worth mentioning because they may be made more likely by action that may otherwise reduce porosity, such as a harsh, low slump mix will have a lower water content or a richer mortar (higher cement/sand ratio) than a sandier mix of equal strength. Obviously a low permeability concrete must be fully compacted by the means available. It must not depend on unrealistic expectations of workmanship. Of course the development of self-compacting concrete is an excellent answer to permeability since it is inherently of low permeability and, at least theoretically, cannot suffer from segregation or a lack of compaction.

Water occupies 15% to 20% of the total volume of fresh concrete and, when the w/c ratio exceeds 0.38 by mass, not all of this water can be consumed in the hydration of the cement. The resultant voids left by the excess water will provide pathways for water transport. If they become discontinuous, they will not provide easy passage for water.

The latest packing theories of mix design have demonstrated that close attention to the packing of fine material of cement size and smaller can reduce total void space in the paste fraction, especially when accompanied by superplasticisers.

The total amount of pore space is not the only factor determining permeability. Another important factor is the distribution of the pores and their discontinuity. Bleeding is a source of continuous or semicontinuous pores. Bleeding is initiated by the settlement of cement particles in the surrounding mixing water, after compaction in place. This tends to leave minute pockets of water under fine aggregate grains. There may be enough water to allow the fine aggregate grains to settle slightly and the water to escape around them and rise up through the concrete. The process occurs on a larger scale under the coarse aggregate particles and eventually the whole mass of the concrete settles slightly, leaving a film of water on the surface. The process can happen very gently without having a great deal of effect on the concrete properties. If bleeding is severe the rising water tends to leave well-defined capillary passages and it is then known as channel bleeding. Water penetration of the hardened concrete is obviously greatly facilitated by both the vertical channels and the voids formed under the coarse aggregate and even fine aggregate particles. It is important to note that concrete does not become "impermeable", as commonly thought, when the capillaries become discontinuous. Discontinuity is a change in the rate of increase in permeability with an increase in capillary porosity as shown in Figure 5.2 (Nokken, 2004).

Reduction of permeability can be effected either by avoiding bleeding in the first place or by blocking the channels after formation. Pore blocking after they have formed takes place as cement continues to hydrate and extends gel formation into the pores. This requires the concrete to be well cured and is greatly affected by w/c ratio. Curing is much more critical for permeability than it is for strength. Another means is to line the pores in the concrete with hydrophobic or pore-blocking material. Such materials are marketed as water resisting admixtures. Hydrophobic materials generally reduce sorptivity more than permeability. Some hydrophobic material may provide an initial benefit but lose its effectiveness in the longer term and so proven materials must be used.

Studies on the effect of drying on permeability by Powers et al. (1954), Vuorinen (1985), and Hearn (1998) showed an increase in the permeability

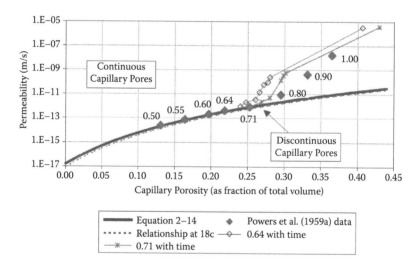

Figure 5.2 Change in permeability due to discontinuity of capillary pores. (After Nokken, M. R., Development of Discontinuous Capillary Porosity in Concrete and Its Influence on Durability, PhD thesis, University of Toronto, Ontario, Canada, 2004.)

coefficient by one to two orders of magnitude when concrete had been dried prior to testing. This highlights that permeability is strongly affected by exposure and resultant microcracking.

5.3.3 Sorptivity

The transport of water in concrete due to surface tension within the capillary network is known as capillary suction or sorptivity. Under conditions of short-term contact with water, the penetration of water is proportional to the square root of time (Ho and Lewis, 1984). The volume of water absorbed by a dry surface will be influenced by the moisture content of the concrete at the time of test. Dhir et al. (1987) compared drying at 20°C, 50°C, and 105°C to develop guidelines for repeatable pretreatment prior to sorptivity testing. They showed that drying concrete with w/cm ratios of 0.4 and 0.55 at 50°C for 14 days removed 50% and 60%, respectively, of the total evaporative water. Even drying at 50°C for 100 days did not achieve an apparent equilibrium in any of the mixes tested as can be seen in Figure 5.3.

Hydrophobic admixtures, controlled permeability formliners, and silane surface treatments all profoundly reduce sorptivity and the reduction compared to an untreated concrete is greater in drier concrete.

5.3.4 Desorptivity

The rate of water loss from an initially saturated concrete surface or desorptivity is also proportional to the square root of time (Dolch and Lovell, 1988). Parrott (1991) concluded that the initial weight loss after four days of uniaxial drying can be regarded as an indicator of the moisture transport properties in the cover concrete.

Baroghel-Bouny et al. (2001) established that isothermal drying of cementitious materials gave a good indication of its permeability. Bentz and Hansen (2000) used x-ray absorption to monitor the effect of drying in cement paste. They tested layered specimens and found that the higher w/c paste dried out first regardless of its location within the composite.

Aldred (2008) found desorptivity was correlated with an apparent steady-state wick action for most concrete types. Therefore a simple desorptivity test, which requires no special conditioning or equipment, appears to provide a good indicator of concrete water transport properties.

5.3.5 Water vapour diffusion

Water vapour diffusion is the movement of water vapour molecules (as a gas not a liquid) due to a concentration gradient and is calculated using Fick's law. Unlike permeability or sorptivity, transport by diffusion is due to random motion of the molecules. Aldred (1999) showed that water vapour diffusion coefficients for concretes incorporating chemical and

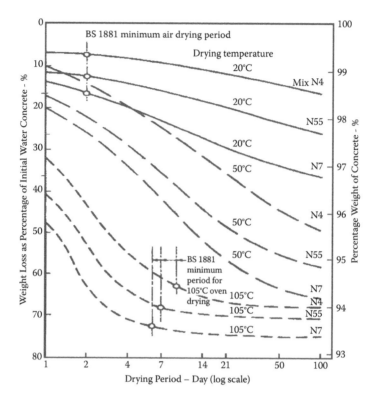

Figure 5.3 Moisture loss under various conditions. (From Dhir, R. K. et al., *Mag. Concrete Res.*, 39, 141, 1987.)

mineral admixtures did not vary as much as other transport properties and appeared unaffected by the initial moisture condition.

Under steady-state conditions, the diffusion coefficient can be calculated using Fick's first law of diffusion:

$$F = -D \times (dc/dx)$$

where
  F = mass flux (kg/m²s)
  D = diffusion coefficient (m²/s)
  c = concentration (kg/m³)

5.3.6 Wick action

Wick action is the transport of water through a concrete element from a face in contact with water to a drying face as occurs in basements, tunnels, slabs on grade, and hollow offshore structures. The term "wick action" to

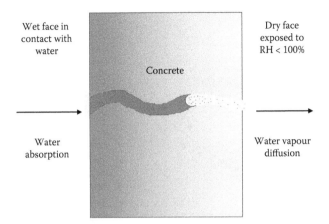

Figure 5.4 Schematic diagram of the interaction of various transport mechanisms during wick action. (From Buenfeld, N. R. et al., in *Chloride Penetration into Concrete*, eds. L. O. Nilsson and J. P. Ollivier, RILEM, Paris, France, 1997.)

describe water transport into air-filled concrete structures appears to have been first used by Aldred (1988). Water transport through concrete due to wick action is many times that due to pressure permeability under typical environmental conditions. Therefore wick action plays an important role in the watertightness and durability of concrete structures (Aldred, 2008).

Wick action was considered a combination of sorptivity and water vapour diffusion with evaporation being the linking process as shown in Figure 5.4. However, Aldred (2008) showed that wick action was poorly correlated to sorptivity but well correlated to desorptivity as shown in Figure 5.5.

James developed a simple equation for estimating steady-state wick action from a simple 14-day desorptivity test:

$$Q_w' = 0.1/L' \times (0.19\, D_{14} - 22.4) \times 10^{-9}$$

where
 Q_w' = estimated steady-state mass flux (kg/m²/s)
 D_{14} = average desorptivity rate over 14 days (kg/m²/s)
 L' = section thickness in metres (dimensionless)

5.3.7 Chloride diffusion

Chloride diffusion is the movement of chloride ions as a result of a concentration gradient. Under steady-state conditions, the diffusion coefficient is usually calculated using Fick's first law of diffusion. Under conditions of uniaxial penetration of chloride ions, diffusion is usually calculated by Fick's second law. However, Fick's law is based on the material through

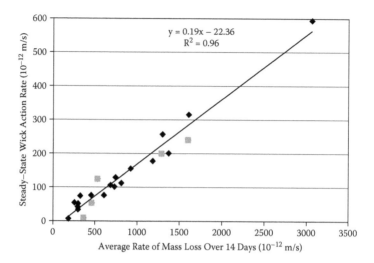

Figure 5.5 Steady-state wick action rates for saturated 100 mm thick specimens compared to average 14-day desorptivity. (From Aldred, J. M., Water Transport due to Wick Action through Concrete, doctor of philosophy, Curtin University of Technology, Kent, Australia, 2008.)

which the ions pass being homogenous and unreactive; concrete is neither. This is one of the reasons why fly ash or GGBS replacement have such a strong influence on calculated chloride diffusion coefficient because penetrating chlorides react with aluminate components.

Under conditions of uniaxial penetration of chloride ions, diffusion is described by Fick's second law:

$$C_x = C_s[1 - erf(l/2\sqrt{D_c\, t})]$$

where
   C_s = surface chloride concentration (kg/m³)
   C_x = chloride concentration at depth x (kg/m³)
   erf = the error function
   l = penetrated depth (m)
   D_c = diffusion coefficient (m²/s)
   t = time of exposure (s)

To estimate the depth of chloride penetration due to diffusion requires an estimate of the surface chloride level. Based on an extensive literature survey, Bamforth and Price (1993) suggested a surface chloride level of 4.5% by weight of binder for predictive purposes for marine-grade structural concrete. Due to chemical binding, fly ash and GGBS usually increase the surface chloride level by approximately 20% more than ordinary Portland cement (OPC) concrete.

A better way to calculate ionic diffusion is the extended Nernst-Planck model, which can be used to describe the flux of multiple ionic species (Samson and Marchand, 1999).

5.3.8 Chloride migration

Chloride migration involves accelerating chloride ion movement by subjecting the concrete specimen to a potential difference. Chloride migration can be tested by both steady-state and non-steady-state methods similar to diffusion. Some concrete technologists dismiss electrical acceleration because it is unnatural. However, ionic movement is necessarily "electrical" because ions are charged particles. Tang and Sorensen (2001) showed that non-steady-state chloride migration (NT Build 492) was directly correlated with the bulk diffusion test, provided the relative maturity was considered.

A fundamental tenet of this book is concrete quality control and timely response to variability. Any rapid and relatively inexpensive test that gives a good indication of an important transport like chloride diffusion is extremely valuable and should not be lightly dismissed.

5.3.9 Resistivity

Resistivity is the resistance of the concrete to the flow of electrical current. It is the reciprocal of conductivity. Resistivity is dependent on the size and tortuosity of the pore system as well as the conductivity of the pore solution. As corrosion of reinforcement is an electrochemical process, its rate will be strongly influenced by the resistivity of the surrounding concrete. Therefore, the resistivity of the concrete not only provides an indication of the penetrability of the concrete but also the rate of corrosion after depassivation has occurred. The most commonly used test for resistivity is the ASTM C1202 or Coulomb test. This is often misleadingly called the rapid chloride permeability test and is discussed in Chapter 7. This standard test takes 6 hours to complete, but the real advantage of resistivity is that it can be measured virtually instantly.

Because resistivity is influenced by both pore structure and the composition of the pore solution, it cannot be used to directly measure either. Pozzolanic materials that react with calcium hydroxide will greatly increase resistivity by both refining the pore structure and reducing the conductivity of the pore solution, particularly highly reactive materials such as silica fume and rice husk ash. Certain admixtures that contain salts, such as calcium nitrite, reduce resistivity primarily by increasing free ions in the pore solution. Hydrophobic admixtures and surface treatments may profoundly increase in situ resistivity by reducing the internal moisture content and the water filled pathways within the concrete. In summary, resistivity is a very important parameter.

5.4 PLASTIC PROPERTIES

5.4.1 Bleeding

Factors affecting bleeding are

1. Amount of fine material (including cement, slag, fly ash, silica fume, and natural pozzolans)
2. Air entrainment
3. Water reduction through admixtures or lower slump
4. Continuity of grading (especially including fine aggregate grading)
5. Use of VMAs (viscosity modifying admixtures)
6. Increased retardation, which delays gel formation and so extends the period during which bleeding can occur

Essentially the fresh mortar in concrete consists of a mixture of saturated solids surrounded by water and any entrapped or entrained air that are lighter than the solids component and will tend to move upward. The better the particles pack together and the more difficult it will be for water to pass through the mass, air bubbles will also tend to block passageways. Cement, slag, fly ash, entrained air, rice husk ash, VMAs, and silica fume (in generally increasing order of effectiveness) are good inhibitors of bleeding. Very fine calcium carbonate (limestone) is a recent development and the superfine material in manufactured sand (crusher fines) is now considered very desirable to control bleeding in some circumstances. Silica fume is the most effective inhibitor of bleeding. It is many times finer than cement and particles of it fill the interstices between the cement particles. Small amounts (as little as 10 to 30 kg per cubic metre) are sufficient to prevent bleeding almost completely. It should be noted that the effectiveness of silica fume is greatly reduced if it is not completely dispersed. Essentially this means that silica fume should always be either batched as a slurry or given adequate mixing time especially together with coarse aggregate to shear the agglomerates and, of course, used with an appropriate superplasticiser.

It should be noted that eliminating or greatly reducing bleeding can create problems with plastic shrinkage cracking. Such concrete may require careful attention to preventative measures such as the use of liquid aliphatic alcohol evaporation retardant (Confilm) or polythene sheeting, mist sprays, and so on.

5.4.2 Rheology

Rheology describes the workability of the concrete. It is a critical feature of most concrete and there is much more to this property than is revealed by the still widely used slump test. Workability testing is more extensively dealt with in Chapter 7. The subject is only briefly covered here.

Apart from slump, workability affects some or all of mobility, fluidity, pumpability, compactability, and, negatively, segregation and bleeding. A factor other than water content is clearly involved and this is best described as cohesion. Cohesion may be physically evaluated in terms of resistance to segregation and bleeding but a numerical measure is needed for use in mix design. Ken Day developed the term MSF (mix suitability factor). This factor is derived from the overall mix specific surface adjusted for the content of cementitious material and entrained air, all of which increase cohesion.

The use of rheometers to measure the yield strength and plastic viscosity of concrete is taking over from traditional testing and traditional characterisation in the laboratory, but their use rarely extends to the field, and these are measured parameters rather than something calculable from gradings and mix proportions. So they are to date a means of establishing whether the desired concrete properties have been achieved rather than a means of calculating how to achieve them, although this may change in future.

MSF is certainly a big advance on characterising mixes only by slump and a verbal description such as pump, structural, or paving mix. However, it is not sufficient alone to cope with the 'new' material, self-compacting or flowing concrete. Even normal pumped concrete needs a measure of grading continuity and bleed resistance. The latter is a matter of having sufficient fine material (at least passing a 200 sieve) or using a suitable chemical admixture such as a VMA.

5.4.2.1 Slump

Although Chapters 8 and 10 use slump as a measure of relative workability, it is important to realise that this is a matter of convenience and that the slump test is a very poor measure of the relative workability of different mixes. One reason for retaining slump as a criterion is that it is so deeply ingrained in the theory and practice of concrete technology. Another is that slump in combination with Day's MSF does have a little more validity as an absolute criterion than slump alone. A third, and probably the most important, is that it is a useful detector of a change in of water content between successive deliveries of the same concrete mix within particular workability ranges.

What is important is not to stop using the slump test, but to realise and allow for its limitations. For example, a limiting slump value is often included in a job specification. With few exceptions, this is not the best way to achieve the specifier's objective. First, there should be an objective for the specification of anything, rather than it having been included in a previous specification and so mindlessly continued in the current document. The objectives may be to avoid high shrinkage, segregation, and bleeding or to avoid an excessive w/c ratio leading to inadequate strength or durability. However, any of these faults can be encountered at almost any slump, however low, and avoided at any slump, however high. It is also easy to detect from a theoretical mix submission, which mixes will be subject to one or other of these problems. The contractor should therefore be permitted to submit his mix for approval at whatever slump he chooses, provided it is designed to accommodate his own slump limit without detriment. It is quite possible to produce fully flowing (250 mm slump or more) concrete having none of the potential faults noted and to produce almost all these faults in a 50 mm slump mix.

5.4.2.2 Self-compacting concrete (SCC)

A whole new ball game in workability has been opened up with the concept of self-compacting concrete, also called self-consolidating concrete or superworkable concrete. This is a relatively new concept, having originated in Japan in the 1980s and originally met with a degree of skepticism in most of the rest of the world. Now it seems quite possible that it will become one of the most widely used kinds of concrete in the not too distant future. This is already true in much of the precast industry.

5.4.3 Pumpability

Bleeding rate (segregation resistance) plays an important role in preventing blockage during pumping. Pumpability should not be considered an intrinsic property of concrete but involves concrete composition, configuration of the plant, and the pumping procedure. Blockages generally occur during priming and restarting after a prolonged delay but not during steady-state pumping. Predicting the expected pumping pressure is a vital part of pumping concrete up super-tall buildings. The pressure required will depend on frictional losses within a concrete pipeline and the pressure head. Kaplan et al. (2005) demonstrated that bends and so on did not influence pressure losses. Pressure calculation is based on Bernoulli's equation. The pressure head component is unavoidable as it is simply a function of the density and vertical height. In the case of the uppermost pour for the Burj Khalifa, the pressure head component alone was over 140 bar. Output (i.e., velocity) and pipe diameter are constructability issues. The critical parameter that will determine the feasibility of pumping to a particular height is the friction factor. This is typically estimated with site trials.

5.5 DIMENSIONAL STABILITY

Dimensional stability may include undesirable degrees of thermal expansion and also disruptive expansion due to alkali–aggregate reaction or sulfate attack but essentially the problem is shrinkage. The major type of shrinkage is thermal shrinkage at early age and drying shrinkage at later age, but there are also autogenous or chemical shrinkage, carbonation shrinkage, elastic defection, and creep under load.

Thermal shrinkage is due to the contraction of concrete as it cools from its peak temperature during hydration to the ambient temperature based on its coefficient of thermal expansion. This is more important in massive elements but can also be an important source of cracking in thinner elements, depending on restraint.

Autogenous shrinkage relates to concretes of very low w/c ratios that shrink as a result of self-desiccation. It occurs much more rapidly than normal drying shrinkage and produces through section shrinkage not from the outside in. Autogenous shrinkage will be additive to thermal shrinkage and is particularly significant in concrete with high replacement levels of GGBS and a low w/cm ratio.

Drying shrinkage is a result of contraction of the cement paste as the uncombined excess water evaporates. This shrinkage is restrained by the aggregates, especially the coarse aggregates. From this it is obvious that shrinkage will be higher if there is more water and cement and more sand. Some coarse aggregates have an appreciable moisture movement that will directly contribute to shrinkage but, apart from this, a higher elastic modulus of the coarse aggregate will reduce shrinkage.

5.6 GOOD APPEARANCE

A good appearance requires that concrete be fully compacted and free from "bug holes". Actually the type of formwork and the mold oil used may have a considerable effect on this aspect.

Bleed control on fair-faced concrete is important. A tendency to bleed allows water to travel up the face of the formwork or toward any slightly leaking joints. This can produce very unsightly results including sand streaks and hydration staining. In its most severe form the latter can result in black areas adjoining joints, caused by the bleed water washing the usual gray dust coating from the cement grains, which are actually black. Since true SCC does not bleed at all, it should be free from such defects. Air voids formed by the release of entrapped air can be a problem, particularly when concrete is cast against inward sloping mold faces. Controlled permeability formliner (CPF) can be helpful in these applications to provide a pathway for removal of air.

5.7 HEAT GENERATION

Heat generation is largely a matter of the type and quantity of cementitious material as well as the level of insulation provided. Low heat Portland cement may or may not be economically available, but in any case it is usually preferable to use a proportion of fly ash or GGBS to reduce generated heat. Where fly ash is not available, some projects have used silica fume to reduce the total cementitious quantity. While the presence of silica fume may result in more rapid heat generation (by speeding the reaction), it permits significant cement reduction at the same strength to reduce total heat generation.

The use of GGBS calls for careful consideration. It actually generates similar or sometimes greater heat than normal cement, but it does so more slowly. So in a typical situation the heat is able to escape and the peak temperature is reduced provided there is not excessive insulation. In massive sections, such as raft slabs more than 3 metres thick, the heat cannot escape quickly enough and the peak temperature may be similar to a pure Portland cement concrete (see Chapter 2, Figure 2.3).

5.8 ECONOMY

The most expensive concrete is that which has to be replaced due to being either initially unsatisfactory or inadequately durable. The cost of a higher quality grade of the concrete itself is, in most cases, a relatively small proportion of the total cost of the final structure. The costs of reinforcement, transportation, placing, finishing, curing, and especially of the formwork, often exceed the basic cost of the concrete. However, it should be borne in mind that the additional cost of a slightly higher quality concrete can be a significant proportion of the concrete producer's profit margin.

The message here is that you should not expect to get any higher quality than you have specified but that it may be worth specifying a quality that is a little higher than the absolute minimum quality you need (Chapter 11, Section 11.2). "Quality" will generally mean a strength grade but shrinkage, bleeding, and resistance to deterioration may need consideration.

Contrary to past practice, the inclusion of cement replacement materials will generally give concrete of improved performance and is often worth specifying rather than merely permitting. At a given strength, the concrete with the lowest cement content will be preferable since it will also have the lowest water content.

Chapter 6

Specification

In the third edition of this book, Ken Day expressed the hope that the practice of mindlessly specifying minimum cement contents and requiring mixes to be submitted and not subsequently varied will have finally died out. However, at the publication of this edition, the practice is still alive. It is certainly not confined to the United States; British and European codes require minimum cementitious contents and maximum water to cementitious materials ratios (w/cm), and transportation departments in Australia also specify particular penetrability performance. Globally most consultants still require mixes to be submitted for approval.

The margin between specified strength and the mean strength required to provide it represents an enormous worldwide expenditure in cost (and greenhouse gas) and billions of dollars are still spent in rectifying deteriorated old concrete and in investigating understrength new concrete. These costs are only reducible by improved technology in concrete production and smarter in specification. Specifiers must have a better understanding of the true requirements of their structure and must ensure the development of better technology in concrete production by allowing producers to profit by it. However it is important not to go overboard in increasing costs by excessive specification detail where not essential.

Chapter 10 of the current edition makes it very clear that the concrete producer must be responsible for designing and controlling concrete mixes, if only because control action must be based on early age results, and taken without waiting for incontrovertible justification. However, there are other reasons. The producer must obtain cooperation from his suppliers and he should be allowed to profit from the development of his expertise. Producers should be encouraged to establish standard mixes and should be allowed to use them wherever possible. Some purveyors of materials, including cement replacement materials and admixtures, and of proprietary mix design and control systems, will be able to offer substantial assistance.

It might be certainly simpler, if you have the necessary knowledge, to specify that particular aggregates, cement and other materials that shall be used in conservative proportions than to specify limits on the properties of all possible materials. However this would require the specifier to assume responsibility for the resulting concrete and would contribute to "dumbing down" the industry and keeping in business less technically competent producers. It would also be likely to substantially increase the cost of the concrete, since a high variability must be assumed, necessitating a large margin over the minimum requirement, having financially disadvantaged the more competent producers. There are certain situations where such a prescriptive specification may be the best option, such as where the local producer has no experience of achieving the required performance, and the precautions that he would add to the concrete mix would be excessive. However, even in this situation, including a performance-based specification as an alternative to the prescriptive one would help encourage producers to develop appropriate mixes and knowledge.

In the past, the concrete industry generally focused on compressive strength. However there are legitimate reasons for specifying more than just a minimum strength. It may be useful to specify a number of requirements in particular cases:

1. A peak temperature and maximum temperature differential to limit thermal cracking or potential delayed ettringite formation
2. A test for reactive aggregates where aggregates without a proven record are considered
3. An air content and spacing factor for freeze–thaw resistance
4. An early strength required for stripping, prestressing, depropping, and so on
5. A drying shrinkage limit
6. A requirement for self-consolidation or extreme pumpability
7. A penetrability test limit for either water penetration or durability in aggressive conditions
8. Maximum crack width for water-retaining structures
9. A bleeding limit, especially where a good off-form finish is required
10. Segregation resistance; it could be specified that the concrete shall not display any tendency to segregation at the proposed workability
11. Abrasion resistance

The problem is that some of these performance goals may be difficult to establish by test at an early age. Some of the test procedures are complex and costly. In addition, the statistical variability of many parameters may not be known and therefore it can be hard to apply the concept of a characteristic performance. Some performance requirements oppose others. The following discussion considers the practical issues associated with performance specification to achieve particular engineering goals. However, having established an acceptable mix, variation from it can be detected by an early-age strength test, even though strength may not be what matters.

6.1 TEMPERATURE RISE

The specifier may want to restrict temperature rise or differential to limit thermal restraint cracking, strength loss, or development of delayed ettringite formation (DEF) in precast or mass concrete applications. Thermal restraint cracking is a common problem. There are two types of thermal restraint: (1) internal restraint due to the temperature differential between the interior and the exterior of the concrete; and (2) external restraint due to thermal shrinkage being restrained by a previously cast element. Most of the significant thermal cracking observed by the authors over the years has been external thermal cracking. However, most specifications focus on the internal thermal restraint and include a limit on temperature differential of 20°C. There are a number of problems with this approach. The value of 20°C relates to the estimated internal restraint for a smooth gravel as shown in Table 6.1 from BS 8110.2-1985, whereas most coarse aggregate used would be crushed and a limit of 27.7°C would be more realistic.

Mass concrete elements are often heavily insulated to achieve this temperature differential limit, which tends to increase the peak temperature and the temperature near the surface increasing the probability and extent of external restraint cracking. As the insulation tends to obstruct the work, the contractor wants to remove it as soon as possible. Therefore as soon as the monitoring of thermocouples is discontinued, the insulation will be removed causing a large thermal differential at the concrete surface and possibly resulting in thermal shock cracking. The authors have found that, except for concrete placed in freezing conditions, ponding with around 50 and 75 mm of water is the great method to facilitate heat loss from a massive element, particularly if it contains large replacements of fly ash or ground-granulated blast-furnace slag (GGBS), as well as preventing thermal shock of the concrete surface. It also ensures excellent curing and helps limit autogenous shrinkage. Some limit on the maximum allowable temperature differential is prudent but needs to consider the type of aggregate used as well as its possible influence on practical construction procedures and peak temperature. One author was involved in the casting of a 15000 m³ self-consolidating concrete raft that took 3 days to cast. With the initial concrete having achieved its peak temperature while fresh concrete was still being placed, clearly the temperature differential within the raft was high but the gradient was low. This is an important point as the temperature differential required to induce cracking is due to the gradient not the absolute differential within a large element. Limiting pour size to achieve some arbitrary differential will generally increase cracking due to additional external restraint.

Specifications sometimes require a peak temperature limit to control external thermal restraint cracking, often around 70°C. However, the peak temperature required to limit cracking will depend on the restraint conditions and the average ambient temperature. In heavily restrained

Table 6.1 Estimated internal restraint for different aggregates

Aggregate type	Thermal expansion coefficient ($10^{-6}/°C$)	Tensile strain capacity (10^{-6})	Limiting temperature drop for varying restraint factor (R)				Limiting temperature differential when R = 0.36
			1.00	0.75	0.50	0.25	
			°C	°C	°C	°C	°C
Gravel	12.0	70	7.3	9.7	14.6	29.2	20.0
Granite	10.0	80	10.0	13.3	20.0	40.0	27.7
Limestone	8.0	90	14.1	18.8	28.2	56.3	39.0
Sintered pfa	7.0	110	19.6	26.2	39.2	78.4	546

Source: BS 8110.2-1985.

elements or cooler environments, a peak temperature of significantly lower than 70°C may be necessary to limit cracking. Modeling is necessary to determine the appropriate value, and guidelines can be found in documents like CIRIA C660. In situations where there is significant restraint for elements with a minimum dimension exceeding around 600 mm or for water restraining structures, probably the best course of action for a specifier is to require that such modeling be done.

Specifications for mass concrete or hot weather concrete usually require a maximum concrete placement temperature, often 32°C but sometimes as low as 20°C. ACI 305 specification recently revised this limit stating that "the maximum allowable fresh concrete temperature shall be limited to 35°C (95°F), unless otherwise specified, or a higher allowable temperature is accepted by architect/engineer, based upon past field experience or preconstruction testing using a concrete mixture similar to one known to have been successfully used at a higher concrete temperature". Although there are important effects of higher placement temperature on plastic properties of concrete, we would suggest that these are the responsibility of the producer in consultation with the contractor. The primary performance goal is limiting external restraint cracking due to the peak temperature. There are different ways of achieving the primary goal. The specified peak temperature for a 4 metre thick raft in Kuwait was 71°C, but the concrete producer did not have access to flake ice to reduce the placement temperature below about 35°C. The solution proposed by one of the authors was to use a high percentage replacement of fly ash. A 55% fly ash replacement achieved the required peak temperature in the raft as well as the strength and other properties. The master specification for the MASDAR development in Abu Dhabi actually prohibited the use of flake ice in the concrete, as the embodied energy involved was contrary to the sustainability goals of the project.

In the case of expected temperature-induced strength loss, temperature will influence strength loss very differently depending on the chemistry of the cementitious binder and therefore the temperature limit should vary depending on the mixture proposed by the producer. Concrete with Portland cement only binder exhibits progressive strength reduction as temperatures increase above about 70°C, whereas concrete containing significant quantities of GGBS or fly ash does not.

Delayed ettringite formation is a form of internal sulfate attack that has been found in heat cured precast elements and in some in-situ mass concrete (Thomas et al., 2008). It is generally accepted that DEF does not occur in concrete when the peak temperature does not exceed 70°C. Guidelines suggest a number of precautions to minimise the risk of destructive DEF in the event of temperatures from 70°C to 85°C.

1. The cement should have a maximum fineness value of 400 m²/kg
2. Portland cement with 1-day mortar strength less than 20 MPa

3. GGBS >35% or Type F Fly ash >20%
4. Silica fume ≥5%
5. Use cement with C_3A between 4% and 10%

Some specifications appear to allow a peak temperature in excess of 70°C on the basis of using supplementary cementitious material (SCMs) in the concrete, which would reduce susceptibility to DEF. However, DEF is a very rare problem in concrete but thermal cracking is not. The default value of 70°C in many specifications may coincide with the generally accepted value above which DEF may occur, but it is also a reasonable value to reduce the thermal restraint cracking and possible strength issues without having to take extraordinary measures.

6.2 ALKALI–SILICA REACTION

Table 6.2 outlines the ASTM procedures to test for alkali–silica reactions (ASR). The procedures range from rapid accelerated procedures such as the ASTM C289 chemical test to longer-term procedures such as ASTM C1293 on concrete, which requires a year or more of exposure. Perhaps the most reliable procedure is petrographic analysis, but this is expensive and samples may not include reactive materials.

Based on the problems with alkali aggregate testing, some authorities have taken the precaution of specifying minimum pozzolanic content in areas where potentially reactive aggregate may be present. Icelandic cement contains 7.5% silica fume to help deal with potentially reactive materials. In Queensland, the use of 20% fly ash is allowed as an alternative to testing for reactivity. Although these are quite good examples of practical prescriptive specifications to overcome a technical problem, problems could arise. The Australian standard for fly ash (AS 3582) does not mention calcium content. Therefore, if a Type C fly ash from a nontraditional source were used, the assumed protection against ASR would be greatly reduced in comparison with the Type F fly ash usually used in Australia. Iceland has its own ferrosilicon plant and a good quality silica fume, however, if it were to import silica flour marketed as silica fume (which is becoming more common), the benefit of 7.5% replacement on ASR would evaporate. Some performance tests are still advisable to protect against poor quality materials or changed circumstances.

6.3 AIR CONTENT

Air entrainment to prevent damage due to freeze–thaw has been widely studied over many years. The use of air entrainment is widely used in North America to reduce freeze–thaw damage. An air content of 6% ± 1%,

Table 6.2 Test methods for alkali–silica reactivity

Test name	Purpose	Type of test	Duration of test	Comments
ASTM C 227, Potential alkali-reactivity of cement–aggregate combinations (mortar-bar method)	To test the susceptibility of cement–aggregate combinations to expansive reactions involving alkalies	Mortar bars stored over water at 37.8°C (100°F) and high relative humidity	Varies: first measurement at 14 days, then 1, 2, 3, 4, 6, 9, and 12 months; every 6 months after that as necessary	Test may not produce significant expansion, especially for carbonate aggregate. Long test duration. Expansions may not be from alkali–aggregate reaction (AAR).
ASTM C 289, Potential alkali–silica reactivity of aggregates	To determine potential reactivity of siliceous aggregates	Sample reacted with alkaline solution at 80°C (176°F).	24 hours	Quick results. Some aggregates give low expansions even though they have high silica content. Not reliable.
ASTM C 294, Constituents of natural mineral aggregates	To give descriptive nomenclature for the more common or important natural minerals; an aid in determining their performance	Visual identification	Short duration; as long as it takes to visually examine the sample	These descriptions are used to characterise naturally occurring minerals that make up common aggregate sources.
ASTM C 295, Petrographic examination of aggregates for concrete	To outline petrographic examination procedures for aggregates; an aid in determining their performance	Visual and microscopic examination of prepared samples; sieve analysis, microscopy, scratch or acid tests	Short duration; visual examination does not involve long test periods	Usually includes optical microscopy. Also may include x-ray diffraction (XRD) analysis, differential thermal analysis, or infrared spectroscopy (see ASTM C 294 for descriptive nomenclature).

(Continued)

Table 6.2 (Continued) Test methods for alkali–silica reactivity

Test name	Purpose	Type of test	Duration of test	Comments
ASTM C 342, Potential volume change of cement–aggregate combinations	To determine the potential alkali–silica reaction (ASR) expansion of cement–aggregate combinations	Mortar bars stored in water at 23°C (73.4°F)	52 weeks	Primarily used for aggregates from Oklahoma, Kansas, Nebraska, and Iowa.
ASTM C 441, Effectiveness of mineral admixtures or GBFS in preventing excessive expansion of concrete due to alkali–silica reaction	To determine effectiveness of supplementary cementing materials in controlling expansion from ASR	Mortar bars—using Pyrex glass as aggregate—stored over water at 37.8°C (100°F) and high relative humidity	Varies: first measurement at 14 days, then 1, 2, 3, 4, 5, 9, and 12 months; every 6 months after that as necessary	Highly reactive artificial aggregate may not represent real aggregate conditions. Pyrex contains alkalies.
ASTM C 856, Petrographic examination of hardened concrete	To outline petrographic examination procedures for hardened concrete; useful in determining condition or performance	Visual (unmagnified) and microscopic examination of prepared samples	Short duration, includes preparation of samples and visual and microscope examination	Specimens can be examined with stereomicroscopes, polarising microscopes, metallographic microscopes, and scanning electron microscope.
ASTM C 856 (AASHTO T 299), Annex uranyl acetate treatment procedure	To identify products of ASR in hardened concrete	Staining of a freshly exposed concrete surface and viewing under UV light	Immediate results	Identifies small amounts of ASR gel whether they cause expansion or not. Opal, a natural aggregate, and carbonated paste can glow—interpret results. Accordingly tests must be supplemented by petrographic examination and physical tests for determining concrete expansion.
Los Alamos staining method (Powers 1999)	To identify products of ASR in hardened concrete.	Staining of a freshly exposed concrete surface with two different reagents.	Immediate results	

Test	Purpose	Method	Duration	Comments
ASTM C 1260 (AASHTO T303), Potential alkali reactivity of aggregates (mortar-bar method)	To test the potential for deleterious alkali–silica reaction of aggregate in mortar bars	Immersion of mortar bars in alkaline solution at 80°C (176°F)	16 days	Very fast alternative to C 227. Useful for slowly reacting aggregates or those that produce expansion late in the reaction.
ASTM C 1293, Determination of length change of concrete due to alkali–silica reaction (concrete prism test)	To determine the potential ASR expansion of cement–aggregate combinations	Concrete prisms stored over water at 38°C (100.4°F)	Varies: first measurement at 7 days, then 28 and 56 days, then 3, 6, 9, and 12 months; every 6 months as after that as necessary	Preferred method of assessment. Best represents the field. Requires long test duration for meaningful results. Use as a supplement to C 227, C 295, C 289, and C 1260. Similar to CSA A23.2-14A.
ASTM C 1567, Potential alkali–silica reactivity of combinations of cementitious materials and aggregate (accelerated mortar-bar method)	To test the potential for deleterious alkali–silica reaction of cementitious materials and aggregate combinations in mortar bars	Immersion of mortar bars in alkaline solution at 80°C (176°F)	16 days	Very fast alternative to C 1293. Allows for evaluation of effectiveness of supplementary cementitious materials.

Source: Farny, J. A., and Kerkhoff, B., *Diagnosis and Control of Alkali-Aggregate Reactions in Concrete,* 15413, Portland Cement Association, Skokie, IL, 2007.

spacing factor of <0.2 mm, and a durability factor of 80% according to ASTM C666 is often specified. Figure 6.1 shows the relationship between air content and durability factor.

ASTM C666 involves 300 freeze–thaw cycles and therefore cannot be used as a compliance test. Until the development of the Air Void Analyser™ (AVA), the only means of determining the spacing factor was the labor-intensive ASTM C457, which involves microscopic examination of a polished specimen of hardened concrete. Accordingly, unless the site has an AVA, the quality control of freeze–thaw resistance has been based on the air content. A recent study was conducted by the U.S. Federal Highway Administration (FHWA) (2006) on the freeze thaw resistance of Vinsol-resin-based air entraining agents compared to a synthetic product. It found that "there are well-established thresholds for the air void parameters that would be expected to give good concrete freeze–thaw resistance. The test data presented in this study suggest these limits may not be applicable in all cases to air entrained concrete containing synthetic admixtures". The fact that changing the chemistry of the air entraining agent had a significant effect on performance but not on the air void parameters believed to be responsible for the performance highlights one of the potential dangers of performance specification. The relationships established with one material may not be valid with another.

Another important point when considering freeze–thaw resistance is that many concrete structures constructed before air entrainment have exhibited good performance over 50 or more years. In Europe, most concrete

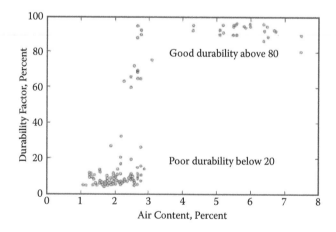

Figure 6.1 Freeze–thaw durability factor for different levels of total air contents. (From Cordon, W.A., and Merrill, D., Requirements for Freezing and Thawing Durability for Concrete, *Proceedings of ASTM* 63, 1026–1036, 1963. With permission.)

is still not air entrained. ASTM C666 (especially Method A—freezing and thawing in water) has been criticised for being too severe, resulting in the rejection of concrete mixes that would have given acceptable field performance. ASTM C666 may be a good test procedure for particularly severe freeze–thaw conditions where concrete may be subject to freezing and thawing in a saturated condition. However, rapid freeze–thaw cycles from all six surfaces of a saturated prism is clearly not representative of the exposure condition of most concrete elements in the field and alternative procedures should be considered.

6.4 EARLY-AGE STRENGTH

Specifiers can use the maturity/equivalent age concept discussed in Section 7.4 together with a thermocouple in the element of interest to estimate the in situ strength as early age. Another good procedure is the use of pullout tests on the actual structure, such as the Lok test that can be used on high-strength concrete. The pullout strength of the hardened in situ concrete can be measured in accordance with ASTM C900 using a system with proven ability to assess concrete with cylinder/cube strength of at least higher than the expected maximum. Systems based on high-strength embedded inserts (such as L-43 inserts and the Lok test) or postinstalled pullout systems (such as the Capo test) can be used. Where the reliability of compliance sampling and testing is in question, the in situ strength of the structural elements should be determined.

6.5 DRYING SHRINKAGE

Specifying a particular drying shrinkage requirement for concrete is quite common. Indeed the Australian standard requires the drying shrinkage at 56 days to be less than 1000 microstrain for all concrete. However, there is trend for specifiers to place more onerous shrinkage limits on concrete. While the drying shrinkage test is an expensive and time-consuming procedure, it is one of nonstrength tests that is regularly used during construction. But how useful is standard shrinkage testing to reduce cracking in concrete structures? As with freeze–thaw testing, the standard shrinkage test is far more severe than most field applications with concrete specimens with a 75 mm × 75 mm cross section drying from all sides into a well-ventilated 50% relative humidity environment. Increasing the dimensions of a concrete element and restricting drying to one or two faces greatly reduces the amount of shrinkage and more important the rate allowing creep effects to reduce the strain (ACI 209, CEB Model Code 1990). A more serious problem with the shrinkage test is the small test specimens that are water cured for 7 days (AS) or 28 days (ASTM) virtually eliminating the effect of autogenous shrinkage as well as making the concrete quality significantly better than would be expected on site. Autogenous shrinkage is a very important component of cracking in higher-strength concrete because it results in rapid through-section strain, which is often combined with thermal shrinkage. As a result of more restrictive shrinkage requirements in an attempt by the specifier to reduce cracking, many producers submit higher strength concrete mixtures that will have lower shrinkage in the standard test procedure but will have higher autogenous shrinkage and often higher thermal shrinkage due to the increased cementitious content. Therefore the concrete may have a greater probability of cracking. If the producer had decided to use a shrinkage-reducing admixture to achieve the specification, he would have reduced both autogenous and drying shrinkage, but it would have been more expensive. Reliance on an unrealistic test procedure or one that does not account for an important factor is always a danger in performance specification.

Another important issue in specifications concerning shrinkage is the value of pour strips to mitigate drying shrinkage in slabs. Pour strips are often proposed to allow the concrete to shrink significantly before closure. As mentioned earlier, virtually all of the shrinkage that will occur in the early curing period is the result of thermal and autogenous shrinkage and this has largely occurred in the first week or two. The different models would suggest a nominal reduction in drying-shrinkage-induced tensile stress even with extended closing time.

The introduction of pour strips interferes with the construction sequence, results in the accumulation of debris, and constitutes a safety hazard on site as well as resulting in an extra construction joint and a strip of relatively poor quality concrete at each pour strip. And at the same time provides no significant reduction in expected drying shrinkage stresses. Accordingly, as long as the construction delay between casting accommodates early thermal (and autogenous) movement, we would suggest that pour strips not be used, particularly for raft slabs where drying shrinkage is greatly reduced due to one-sided drying and the thickness of the concrete element.

6.6 RHEOLOGY AND SELF-CONSOLIDATING CONCRETE

There is a tendency to limit concrete workability in specifications based on the assumption that lower workability produces better concrete. Although often true when added water was the only way to increase workability, it is certainly not true in the age of advanced admixtures. Poor workability can lead to honeycombing, slower construction, and uncontrolled water addition after compliance sampling. Resultant defects can lead to costly repairs and litigation where the specification will come under scrutiny. The problem of prescriptive specification of rheology can also occur with self-consolidating concrete (SCC) where overzealous specifiers can require very high workability parameters that can lead to segregation. We would suggest that the specification require that the contractor or premix company confirm that the rheology of the concrete is satisfactory for the proposed placement procedure and the mix developed complies with the performance parameters. Assessment that the rheology of the mix is acceptable on site should be based on the supplier's proposed value and established tolerances.

There is a tendency to want to specify a w/c ratio, since this is the best overall criterion of concrete quality. There are three reasons not to do this. One is that strength is much easier to use as a control. Another is that if there is some factor causing a departure from the anticipated w/c versus strength relationship (such as bond to coarse aggregate), then strength is the better guide. Third, the necessary average quality will be dependent on the degree of control and producers able to achieve lower variability must be allowed to profit by it. The use of tests such as the AASHTO T318 microwave test to check fresh water content is an excellent tool for producers to control the water content of their concrete, but we question its value as a compliance test to confirm w/c ratio, although the Port Authority of New York and New Jersey (PNY&NJ) seems to have had success with using the procedure for this purpose.

6.7 DURABILITY

Premature deterioration of reinforced concrete due to chloride-induced corrosion of reinforcement is a global problem that costs billions of dollars annually. In severe environments, concrete structures have often failed to achieve their required service life without major maintenance. As more specifications now require a minimum design life of 100 years or more for major projects and infrastructure, there is even more demand for appropriate specifications to ensure the durability of reinforced concrete. International codes provide prescriptive solutions to increase the required concrete quality and cover thickness to improve chloride resistance.

Following is a brief summary of the durability guidance given in various codes and standards for concrete exposed to a marine environment.

Australian Standard 4997 "Guidelines for the design of maritime structures" classifies the splash zone as C2 and recommends minimum concrete strength of 40 MPa with a minimum cover of 75 mm to achieve a design life of only 25 years. The notes list a number of additional measures to achieve a longer design life. AS 3600 (Concrete code) provides guidance for a service life of 40 to 60 years. In AS 3600-2009 exposure to the tidal and splash zone would be classified as C2 exposure and require a minimum 50 MPa compressive strength and 65 mm cover.

ACI 318-11 deems exposure to moisture and an external source of chlorides as severe and requires a maximum w/cm of 0.40 and a minimum characteristic strength of 5000 psi (34.5 MPa cylinder strength).

British Standard (BS) 8110: Part 1: 1997 provides no explicit reference to the intended design life. Regarding durability requirements, concrete in seawater tidal zone down to 1 m below the lowest water level would be classified as "most severe". The minimum requirements are C50, maximum w/cm ratio = 0.45, minimum cementitious content of 400 kg/m³, and minimum cover of 50 mm.

BS 6349: Part 1: 2000 classifies concrete structures in the upper tidal and "dry" internal faces of submerged structures as XS3. Limiting values for composition and properties of concrete for a required design working life of 100 years in UK seawater conditions are 37 MPa (cube strength), maximum w/cm ratio = 0.45, minimum cementitious content of 370 kg/m³ (based on 50% < ground granulated blast-furnace slag (GGBS) < 80% or 35% < fly ash (FA) < 55%) and minimum cover of 50 mm.

BS 8500-1:2002 also classifies concrete in tidal zone to be classified as XS3. Limiting values for composition and properties of concrete for a required design working life of 50 years in UK seawater conditions are C45, maximum w/cm ratio = 0.40, minimum cementitious content of 360 kg/m³ (based on 36% < GGBS < 65% or 21% < FA < 35%), and minimum cover of 40 mm. Interestingly, the requirements for 100-year design life are not given.

Building Research Establishment Special Digest No. 1 (SD1) was written primarily in response to the growing recognition of the occurrence of thaumasite form of sulfate attack (TSA) in UK buildings and structures since the 1990s. Thaumasite is a form of sulfate attack that only occurs at cold temperature. While not related to marine environments, SD1 provides useful guidance on the specification of concrete for installation in aggressive ground conditions. The guidelines for concrete specification include recommendation for 100-year design life. Based on a hydraulic gradient less than 5, Table E1 recommends DC-4 and an additional protective measure (APM) for an intended working life of at least 100 years. This equates to a concrete mix containing minimum 380 kg/m³ cementitious incorporating 36%–65% GGBS or 21%–35% FA, maximum water/cement ratio of 0.40. A waterproofing membrane, controlled permeability formliner or others would be considered the additional protective measure.

Concrete Society 163 "Guide to the design of concrete structures in the Arabian Peninsula" was published in 2008 and would classify all of the elements in contact with the ground and up to 3 metres above as extreme. For an intended life of only 30 years, the recommended mix is C60, maximum w/cm = 0.35 and minimum cementitious content is 400 kg/m^3 for a 70 mm nominal cover. This is based on 66%–75% GGBS or 36%–40% FA or a ternary blend. The guide also suggests that special structures with extended design lives would require special consideration and may need enhanced protection, such as admixtures, corrosion resistant rebar, surface treatment, or electrochemical methods. This guideline does advocate service life modeling to refine recommendations.

Unlike the Australian and American codes, the European codes and guidelines focus on the composition of the cementitious materials used to achieve durability. Although the authors would agree that fly ash, GGBS, and other supplementary cementing materials generally have a significant beneficial effect on durability, there is considerable variation among different commercially available products. Of particular relevance to the chloride penetration is the aluminate component, which strongly influences the chloride binding capacity. A particular problem encountered recently has been so-called silica fume with a high silicon dioxide content but low pozzolanicity due to the presence of large quantities of crystalline silica.

Therefore total reliance on a requirement for particular levels of cement replacement based on assumed qualities of the materials has some risks in terms of durability. In addition, confirming the quantity of GGBS, FA, or silica fume in a hardened concrete sample is also difficult and expensive in the event of dispute about the quantity used. This has not been considered a potential problem in most specifications requiring fly ash and GGBS because they were considered less expensive. The reduced availability of good products does make this assumption less likely.

In addition to the code requirements, there has been an increasing trend to specify performance limits based on different transport properties. Chlorides can penetrate concrete through capillary absorption into unsaturated concrete, wick action, and diffusion through water filled pathways within the matrix driven by a concentration gradient. However, unlike compressive strength, there is little information available on the expected variation in the results obtained as well as on the relationship between such compliance tests and in situ properties/performance. Indeed, unlike air entrainment to enhance freeze–thaw resistance, the required performance for the different specified parameters to achieve the desired durability has often not been established.

Faced with absolute performance limits suppliers have tended to significantly overdesign their concrete mixtures to help ensure compliance, which increases production cost with unknown benefit in terms of durability enhancement. The use of additional cementitious material to achieve certain performance limits at early ages may sometimes have a detrimental effect on fresh and hardened properties and increase the environmental impact of the concrete.

An unexpected consequence of the increase in performance specification has been the submission of inappropriate concrete mixtures just because the necessary test data is available so that the producer did not have to conduct additional trial mixes and long/expensive testing. This would be analogous to requiring a range of complex tests such as creep on all concrete mixes. The concrete industry came to accept that compressive strength was a key parameter and easy to measure, and this provided an indicator of other mechanical and deformation properties. Based on the work by Ho and colleagues, the Australian standard has tried to use compressive strength as an indicator of durability as well but that has been a bridge too far.

What we need is more real field data on the actual performance of concrete in aggressive environments related to their early age properties as well as further work on simple and inexpensive early-age tests for chloride penetrability and water transport. There have been many test procedures proposed for this purpose. We would suggest that the best contenders would be absorption for physical salt attack, desorption for water transport, and chloride migration for chloride penetration with a simple resistivity test providing the continued compliance test. The STADIUM program from North America has done a good job of relating field performance to early age properties producing arguably the most comprehensive service life prediction model available. The early-age testing involves permeable voids, desorption, and chloride migration coupled with petrographic analysis to confirm chemical properties. However, there are still assumptions regarding the expected exposure conditions and improvement of penetrability with time, but it is a step in the right direction. Not all projects are going to conduct a detailed assessment of service life but simple/cheap compliance tests based on resistivity and desorptivity could easily be added to compressive strength to provide much more information on the concrete's potential durability. When tests are cheap and simple, accumulating statistical data is easy and producers would be encouraged to get to understand how to optimise their mixes rather than the current situation of sticking to a mix because it has a compliant diffusion coefficient.

An appropriate performance-based specification for durability in a chloride environment could be based on chloride migration testing over a period up to 90 days to estimate the chloride penetration resistance and the improvement over time during mix development. Chloride migration testing can be correlated with a simple and rapid resistivity test, which could be used during production for ongoing quality control.

6.8 CRACK WIDTH

Specification of maximum crack width is a very contentious subject. Cracking in reinforced concrete structures is complicated. This is the reason that CIRIA C660 defines "maximum" crack widths at the surface of the concrete as design "target" characteristic values with only a 5% chance of being exceeded. Using this probabilistic approach to cracking would solve a number of disputes regarding crack widths. Crack size is limited in water retaining structures to enable autogenous healing and crack size is also limited in an attempt to improve durability. We would agree with the former but question the latter. Clearly cracks along the reinforcing such as plastic settlement cracks are a serious durability issue regardless of their size. However, if transverse cracks are important to durability, why increase their number by adding more reinforcement to disperse them as pointed out by Beeby in the 1980s.

BS8007 advised that self-healing should seal cracks of 0.1 mm in 7 days and 0.2 mm in 21 days. AS3735 Liquid Retaining Structures does not nominate a limiting crack width but mentions leakage at a crack in section 7.3 Testing of Liquid Retaining Structures. This section implies a limit on crack size at the time of testing equivalent to tightness class 2 in EN1992-3 Liquid Retaining and Containment Structures.

EN1992-3 Liquid Retaining and Containment Structures acknowledges that cracks tend to self-heal. The recommended maximum surface crack width is 0.2 mm, where the ratio of wall thickness to hydrostatic pressure is less than 5, reducing to 0.05 mm, where that ratio is greater than 35. EN1992-1-1 Design of Concrete Structures provides guidance on durability considerations. Table 7.1N provides recommended values of $w_{max} = 0.3$ mm for exposure class XS3, which is defined in table 4.1 as corrosion induced by seawater in tidal and splash zones. Therefore, a crack width of 0.3 mm should be considered adequate to achieve the required durability against aggressive conditions even if full self-healing has not occurred within 3 months. The provisions for crack size from AS3735 and EN1992-3 apply at the time of test or at the time of filling with water if there is no test.

Mohammed and Hamada (2003) examined the corrosion of reinforcement passing through a construction joint (i.e., a man-made crack). They found that autogenous healing at the joint plane prevented corrosion except where the joint had been treated with epoxy and latex paste. Calder and Thomson (1988) report significant corrosion in cracks that had been fully filled with epoxy. We would suggest that a crack width limit of 0.30 mm at the concrete surface is appropriate in most chloride environments

6.9 DEVELOPMENT OF STANDARD MIXES

Specifications have tended to assume that the concrete supplier will design a special mix to comply with the specification. This may be necessary in relatively rare cases, but it does have some disadvantages:

1. No history of previous satisfactory performance on actual projects.
2. No common pool of test results with same mix on other projects.
3. Truck drivers less familiar with mix, less able to judge workability and detect abnormality.
4. Variability may be increased if every now and then the standard mix is supplied in error.

It might be reasonable to provide a financial advantage to suppliers who have satisfactory standard mixes in use, under routine control and with a range of properties established. The form of encouragement could be to allow a reduced testing frequency for such mixes and to require pretesting, and a higher testing frequency for the first months, of new mixes.

The aforementioned points apply even for major projects, but their importance is far greater for the many "ordinary" projects that probably account for most of concrete produced. Small projects cannot economically generate sufficient test data to maintain good control. This means that they are essentially dependent upon the producer's quality assurance system. In such circumstances it is counterproductive to specify nonstandard mixes unless absolutely essential. It is possible that a very small project could nevertheless derive great advantage from the use of 100+ MPa concrete in a particular column or involve a single wall of exposed aggregate concrete of super critical appearance. In such circumstances special mixes are obviously involved and control costs are of little importance. However, a refusal to accept a standard mix for a 25 MPa internal floor slab would be justified only if the standard mix were distinctly unsatisfactory.

The specifier should generally concentrate on obtaining full information, both past and current, about standard mixes. The aim should be to check that the supplier's control system is working well rather than to supplant it. These remarks are relevant when only compressive strength is regarded as important. This chapter deals with requirements other than strength, and the importance of using standard mixes of established performance is much greater in respect of such requirements.

A time is coming when it may be less essential to use standard mixes. The control system pioneered by Day enables results from many grades to be combined onto a single control graph. The performance of mixes may be seen in terms of factors in mix design equations rather than a stand-alone assessment. The same situation has been encountered in many different industries (Toffler, 1981). Initially, mass production requires acceptance of a reduced range of products. However, as the technology of both production and quality control advance, the standardisation necessary tends to be that of small parts of the whole. In this way products of very wide variety can be produced from components that are rigidly standardised. It is emphasised that this stage has not yet been reached in concrete technology and specifiers should currently concentrate on the second phase of reduced variety. However Day presented a paper "Just-in-Time Mix Design" at ACI Cancun in 2002 that demonstrated the necessary technique for this development. This was further referred to in his paper "Concrete in the 22nd Century" at the CIA Biennial in Melbourne, October 2005 (Day, 2005b).

6.10 BATCH PLANT EQUIPMENT

The availability of computer-operated batching equipment, able to positively record the actual as-batched quantities for each batch of concrete, is an important factor in the control process. It provides the following advantages:

1. It gives a considerable degree of assurance that the batches sampled are in fact typical of the whole output. This greatly strengthens the argument in favor of a reduced rate of testing, allowing emphasis on quality of testing, and a thorough analysis of the results rather than sheer volume of testing.
2. It provides a ready means of adjusting mixes and of keeping accurate records of what adjustments were made and when.

It is therefore fully justified to specify that such equipment should be used on any important work and that the resulting databank should be made available to the supervising team. Should such equipment not be made mandatory, it would be reasonable to halve the otherwise envisaged sampling rate if it were provided.

6.11 PROPOSAL–APPROVAL SPECIFICATIONS

Without increasing cost excessively, it is virtually impossible to so specify a concrete mix that it will necessarily be satisfactory. Strength, slump and surface area (as measured by Day's mix suitability factor [MSF]) can be specified but problems can still result from details of the combined grading. Mix design should be a matter of combining available materials so as to minimise any disadvantages they may have individually. It is possible to specify conformance of each individual material to ideal requirements so that they can be combined in standardised proportions, but this is usually only practicable on large contracts for which aggregates are being specially produced. Even so some variation is inevitable, and it is difficult both to require rigid compliance with specified proportions and to provide for variation. This path leads to full acceptance of total responsibility for concrete quality by the supervising authority, which is undesirable for many reasons (from needing to take over control of incoming materials quality to facing claims by the contractor that any defects in the finished product are due to matters beyond his control). The Australian government airfield construction branch used such techniques in the 1980s. Excellent concrete resulted, and it was considered by those in charge that the high cost was justified by the importance of the work.

The preferable course is to specify as closely as possible the properties required of the concrete and require the contractor to set out in full detail exactly how he proposes to provide them, including his specification limits on incoming materials and within what limits and to what accuracy he proposes to adjust the mix. This clearly gives the contractor absolute freedom to propose the most economical and practicable way of providing concrete of the required properties. It is much easier to detect any unsatisfactory features of such a proposal than it is to so specify a mix that it could not possibly have any unsatisfactory features.

Once the contractor's proposals have been accepted by the supervisor, they become the specification. Insistence on conformance to this specification is easier since the contractor, having proposed it himself, cannot claim it to be unrealistic in any way, and there can be no surprise "loopholes" in the original specification. Of course, in the authors' opinion, even this type of individual attention to mix regulation by a purchaser would only be justified on very large projects, usually those with a dedicated supplying plant.

6.12 SHOULD MIXES BE SUBMITTED?

An important question is whether mixes should be submitted for approval, and if so, approval by whom. It seems reasonable that a purchaser should be entitled to know what is in the concrete he is purchasing. The purpose of such a submission should be to ensure that the mix has no objectionable features. These might include admixtures containing calcium chloride, air-entraining agents known to give an excessive bubble size, potentially reactive aggregates, and aggregates known to have high moisture movement or to cause popouts in exposed surfaces. The list is not extensive and a list of materials rather than mix proportions might meet the need, however, to be effective assessment needs to be by a qualified and experienced concrete technologist.

Militating against detailed mix submissions is the desirability of using standard, well-proven mixes from the viewpoint of quality control and a proper degree of confidentiality from competitors. Also the producer needs to be entitled to vary his mixes within limits from day to day to maintain control.

6.13 CASH PENALTIES

It has been pointed out that the producer must be allowed to regulate the mix on the basis of early age data and compliance with requirements should be assessed on the statistical analysis of a substantial number of later age results, for example, the last month's or the last 30 results, whichever is larger. The important point is that a distinction be made between results that are below specification requirements and concrete that is unacceptable and must be replaced, strengthened, or treated in some way. If any concrete is unacceptable, this cannot be assumed to be unique and all concrete of the same grade in the same period must be examined. It is not satisfactory to attempt to locate and core concrete that has given a low result assuming that untested truckloads will be acceptable.

Strength results are assumed to be normally distributed and their mean strength is required to be 1.65 SD or 1.28 SD above the specified strength according to whether a 5% or 10% defective criterion has been specified. Only 1/1000 results are theoretically expected to be more than 3.09 SD below the mean and only 1/100 below 2.33 SD. Essentially this means that there is a negligible likelihood of a result more than 1 SD below the specified strength if the concrete is acceptable, but concrete up to 1 SD, or say 5 MPa, below the specified strength cannot be considered to be unsafe.

So the proposal is that concrete failing the specified limit but with (statistically) no more than 5% more than 5 MPa below the specified strength be accepted with a cash penalty of say 1% of X truck price per 0.1 MPa shortfall (so concrete marginally inside the acceptance with penalty limit will incur a penalty of almost 5/0.1 = 50%). This should be based on a statistical analysis of the last month's or the last 30 results, and apply to all the concrete of the grade in question supplied in the period.

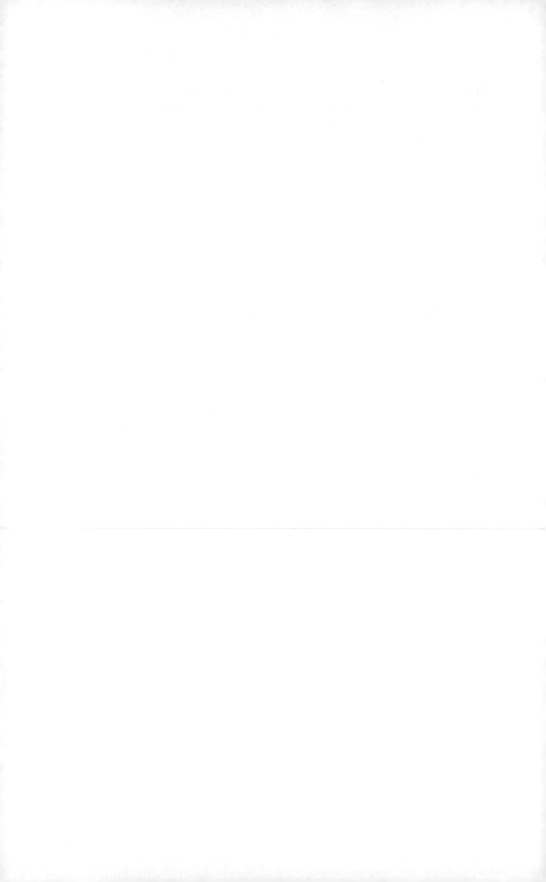

Chapter 7

Testing

7.1 PHILOSOPHY OF TESTING

It is very important to understand the philosophy of testing. Only people ignorant of the true situation regard a test result as an accurate portrayal of the property tested. Unfortunately this tends to include many people in authority such as specifiers, controllers, and legal people.

First, no test can be perfectly accurate and it is as well to consider how inaccurate it might be. Second, the sample tested may not be truly representative of the mass being assessed. For example, standards may require great care in checking the equipment and following a rigid procedure to get an accurate sand grading. It may also present clear rules for obtaining a representative sample. But if you are doing quality control (QC) on concrete, there is nothing better than doing frequent rough checks (twice the number of tests in half the time), looking at the results on a cusum graph of specific surface, and taking a second sample to confirm if the first one says there has been a change. Another important point is that it is good to test different parameters whenever practical so that aberrant results can be weeded out and true change points confirmed by more than one parameter. On one major project in the Middle East, the compressive strength reduced at the same time as the resistivity. Coincidentally the huge building boom at the time had resulted in a shortage of silica fume that was in the design mix. The premix company was accustomed to arguing about the vagaries of compressive strength testing but had no answer to two parameters simultaneously changing.

A very important distinction between QC and research is continuity. A research project, however large and long, must eventually come to an end, and some very elaborate statistical techniques and great care to achieve testing accuracy may be of substantial value in reaching an accurate conclusion. QC is a continuing flow of data that may necessitate revised conclusions from time to time. Many factors may affect the desirable level of sophistication of both testing and analysis techniques. The cost benefit must be assessed of the relativities of expense and accuracy against volume and simplicity, especially taking into account the standard of personnel who will be operating the system.

A vital component of the shift from prescriptive specifications to performance specifications advocated in this book is the reliability of testing. Areas where specifiers do not trust the testing houses to give valid and accurate results will depend on prescription because they feel safer. The establishment of the recognised National Association of Testing Authorities (NATA) accreditation scheme in Australia was a crucial part in the general acceptance of performance specification. Infamous cases, such as where a grand jury charged the largest materials testing laboratory in New York with systematically falsifying results, do not help build confidence.

As with the rest of this book, the authors strive for truth and reality over regulation and convention, but warn readers that there may be times when their sometimes unconventional views are unacceptable to someone who has to be humored. It is possible that we may be the ones who are wrong.

7.2 RANGE OF TESTS

A very large number of tests on concrete have been devised. Following is a partial list.

Tests on hardened concrete
    Compressive strength (cylinder, cube, core)
    Tensile strength
        Direct tension
        Modulus of rupture
        Indirect (splitting)
    Density
    Shrinkage
    Creep
    Modulus of elasticity
    Absorption
    Permeability
    Resistivity
    Freeze–thaw resistance
    Resistance to aggressive chemicals
    Resistance to abrasion
    Bond to reinforcement

Analysis for cement content and proportions
In situ tests
    Schmidt hammer, pull-out, break-off, cones, etc.
    Ultrasonic, ground-penetrating radar (GPR), Impact-Echo, nuclear, resistivity
Heat generation
Tests on fresh concrete
    Workability (slump and over 20 others)
    Bleeding
    Air content
    Setting time
    Segregation resistance
    Unit weight
    Wet analysis
Temperature

Of these many possible tests, in practice well over 90% of all routine tests on concrete are concentrated on compression tests and slump tests that should be, but are not always, accompanied by fresh concrete temperature and hardened density determinations.

Before considering whether this is a desirable state of affairs, it is first necessary to consider the purpose and significance of the testing. There are at least three possible purposes:

1. To establish whether the concrete has attained a sufficient maturity (for stripping, stressing, depropping, opening to traffic, etc.)
2. To establish whether the concrete is basically satisfactory for the purpose intended
3. To detect quality variations in the concrete being supplied to a given specification

It is very important to be clear about the purpose of the testing because attempts to fulfill all these purposes simultaneously usually lead to inefficiency in fulfilling any of them. The true purpose of the majority of tests is the detection of quality variations.

The selection of compressive strength for the great majority of control testing relies upon three basic assumptions:

1. That most other properties of concrete are related to compressive strength
2. That compressive strength is the easiest (most established), most economical, or most accurately determinable variable amenable to test
3. That compressive strength testing is the best means available to determine the variability of concrete.

The second of these assumptions will be examined in detail later.

The first assumption is probably correct in so far as the purpose of the test is to detect quality variations but is not necessarily correct if the purpose is to establish whether the concrete is basically satisfactory based on non–strength-related performance parameters such as freeze–thaw resistance, shrinkage, or penetrability, which will be discussed later in this chapter.

It may well be impracticable on most projects to use other forms of testing for quality control purposes, although rapid wet analysis and simple resistivity testing has been so used. However, especially when we are dealing with standard mixes from a premix plant or a special mix designed for a specific purpose, it is certainly practicable to carry out a much wider range of tests to initially verify a new mix design and to repeat a wide range of tests at say annual, or six monthly, intervals for standard mixes. An excellent example of this is the shrinkage of concrete in the Melbourne, Australia. For many years structural designers had been concerned about excessive shrinkage but the only action resulting from this concern was to prohibit the use of pumped concrete on some projects and limit sand percentages on others. However, in 1977–1978 CSIRO (the Australian Government Commonwealth Scientific and Industrial Research Organisation) carried out shrinkage tests on a range of standard Melbourne area pump mixes and showed a wide range of variation with clearly definable causes. It then became practicable to specify a limiting shrinkage and in most cases to permit the use of pumped concrete since the tests showed that some pumped mixes had a lower shrinkage than some nonpump mixes (the factor involved being the influence of the coarse aggregate).

Similar action is now needed in respect of splitting strength, permeability, durability, abrasion resistance and also workability (other than slump), segregation resistance, bleeding, and surface finish characteristics. These were all matters on which we were flying as blind as we used to be on shrinkage at the time of writing the first edition. In the intervening years there has certainly been substantial action in respect of durability and penetrability (with the latter seen as the best available criterion of the former).

7.3 COMPRESSION TESTING

Considering now the accuracy and convenience of compressive strength as a routine control, the situation is not as simple as was thought 20 or 30 years ago. In Australia we are fortunate to have the world's first and most highly developed NATA. We have a better system than most other countries for ensuring that test specimens are cast by competent people, taken to laboratories with satisfactory curing facilities, capped with a sound cap, and tested in a standard manner in a properly calibrated and maintained testing machine. Without being able to quote chapter and verse, but having used both extensively, the authors also have come to the view that a cylinder specimen is at least a little more reliable than the cube specimen. Nevertheless it has now apparent that NATA certification is not sufficient to ensure that different laboratories obtain essentially the same test strength on concrete from the same truck of concrete. Isolated differences of over 10 MPa and consistent differences of the order of 2 to 4 MPa have been documented in Melbourne (Day, 1979, 1989). There are two aspects to the problem:

1. The technology of compression testing machines
2. Day-to-day performance variation

7.3.1 Testing machines

A compression testing machine is usually by far the most expensive item in a routine concrete QC laboratory. As such machines are also very durable items, there is a tendency for quite antique versions to be still in service (and indeed they may give better results than a cheap new machine).

It is apparently a simple thing to apply a compressive load to a test specimen using a hydraulic ram. However, in practice it is far from simple because the results obtained must be very consistent and must bear comparison with other testing machines.

Ken Day has had a wide experience of operating different classes of compression testing machine over many years, but such general experience is of little value. What matters is access to comparative results on samples from the same truck of concrete and preferably cast by the same person. A requirement that this be done as a regular routine has been part of the Day's standard specification for some years and such data is therefore available covering a number of different pairs of laboratories. The Australian NATA also organises occasional comparative tests in which a large number of specimens are cast from a single truck of concrete and distributed to many laboratories. There is a distinct difference in the extent of variation found when each laboratory is on its mettle in a major isolated comparative exercise and that found when the comparison is under everyday routine conditions.

A 2 MPa strength difference is equivalent to a cement content difference of between 10 and 20 kg/m³ (17 to 34 lb/cu yd). A single testing laboratory may well be controlling a production of 10,000 to 100,000 cubic metres of concrete per month (from several plants). So that "high" cost of a testing machine may be little more than the difference in the cost of cement requirement according to two different machines *per month*.

7.3.2 Testing machine technology

Obviously a correct result will not be obtained unless the stress is uniformly distributed over the test specimen (and any deviation in this respect will lead to a lower result).

An assumption is made that the faces of both the test specimen and the testing machine platen are absolutely flat and that the load will be applied concentrically. Quite small differences in planarity can make very large differences in contact area and therefore in stress distribution. With cube specimens this problem will worsen with older and higher strength specimens because the older concrete (i.e., 28 day rather than 7 day) will be more rigid, that is less subject to plastic distortion. With cylinders the problem is different. Here the capping compound (e.g., where sulfur caps are used) will flow equally at any age. The platen planarity may be slightly less critical but any plastic flow allows stress concentrations to develop unless the original cylinder ends are very close to flat.

Spherical seatings are provided to allow one platen to rotate to compensate for any tendency for the two opposite faces of the test specimens not to be exactly parallel. This introduces its own problem in that, if the spherical seating were effective during the whole test, any eccentricity at all would lead to a bending moment in addition to an axial force, so reducing the failure load. Therefore spherical seatings must be lubricated with a very light machine oil specifically so that the oil will break down under pressure and allow the seating to lock solid after an initial adjustment. Extreme pressure lubricants, such as graphite grease, must be avoided, as they will produce lower and more variable results. For cubes this is even more important because, since the specimen is tested perpendicular to the direction of casting (and therefore water gain or bleeding), its physical center may not be its "center of resistance", that is, if the cube is stronger at the bottom than at the top, its center of resistance would be displaced toward the previously bottom face when turned on its side for testing.

A further influence of the platen–specimen interface, again especially with cubes, is that friction provides a lateral restraint to the Poisson's ratio spreading effect and so increases the test strength. Day (inadvertently) demonstrated this many years ago when he tested cubes coated with a wax curing compound. The compound may have increased the actual concrete strength, but it certainly caused a drastically reduced load at failure. The reason for test cylinders to have a height to diameter ratio of 2 is to avoid this effect in the central area where failure actually takes place. This is probably the main reason for the difference between the test strength of cubes and cylinders from the same concrete. It may also be the reason why this effect is reduced at higher strengths. However, a further reason is that bleeding voids, which are more likely at lower strengths, may have a greater effect on cubes than cylinders owing to the different orientation during testing.

7.3.3 Bad concrete or bad testing?

Day was invited to give a paper on the aforementioned topic to the 1989 ACI San Diego Convention (Day, 1989). The paper has not been published (it is however now on Day's website), but the conclusions presented and the fact that an ACI session organiser requested a paper on this topic indicate that the question merits close attention.

The first half of the paper presented factual data showing that it is unreasonable to expect that a properly presented result from a reputable testing laboratory will always be an accurate representation of the quality of the concrete. Pair differences exceeding 5 MPa were noted for apparently identical test specimens from the same truck of concrete tested by the same laboratory. Seven- to 28-day strength gains were also shown to be capable of ±50% from sample to sample of concrete of the same mix design using the same materials. The clear conclusion was that a strength test result was a totally unreliable piece of information. The audience awaited Day's proposal of some more satisfactory means of assessing concrete quality than a compression test.

The second half of the presentation showed that the very same data used in the first half could be analysed to show quite accurately when a genuine change in concrete quality occurred. Cusum graphs of 7- and 28-day strength showed downturns and upturns on exactly the same dates in spite of individual differences. The two laboratories showing the large differences on individual samples nevertheless agreed exactly as to when these change points occurred.

The overall conclusion presented was that an appropriate analysis of a series of test results can yield very reliable conclusions but that any individual test result should be regarded with great suspicion. Some of other conclusions presented were as follows:

1. Concrete producers are not so good that it is unnecessary to test concrete nor testing labs so bad that it is ineffective to do so.
2. There is no better complete replacement for traditional cylinder/cube testing because it is the only way in which the combined effects of batch quantity variation, material quality variation, silt and dust content variation, air content and temperature variations, delivery delays, and added water effects can be integrated.
3. We must cease to think of a single test result as an invariably accurate judgment as to whether a particular truck of concrete is acceptable. First, it may well not be accurate, and, second, we should show as much concern for those trucks we did not test as for those we did test.

Rather we should regard the analysed pattern of test results as an important part (but only part) of the evidence we require to establish whether the totality of concrete being delivered to the project (or leaving the plant) is or is not of the required quality.

4. Before concrete of a particular grade is even ordered, it should be established that it is almost certain to be satisfactory. This may be done on the basis of trial deliveries, laboratory trials, analysis of past data, or even just the reputation of the supplier. This assessment needs to take into account variability as well as mean strength. For an important project it may be inadvisable to obtain concrete from a supplier who cannot show either or both substantial analyses of past data showing low variability or a computer batching plant that records the actual batched weights of every truck load delivered.

5. A particular individual (perhaps with assistants on a major or widely spread project) should have the responsibility of visually inspecting every truck of concrete and rejecting or further testing any suspect loads.

6. When a truck is sampled and test specimens cast, there should normally be at least three specimens. This is to permit an early-age test and a pair of 28-day tests although it is better to have a pair of results for the early age. The early age (not later than 7 days) is because any necessary mix adjustments must be carried out long before 28-day tests are carried out. The 28-day test is necessary to establish the current significance of the early age results. Two 28-day specimens are needed partly because the average pair difference is the best measure of testing quality and partly so that one can be brought forward to confirm or amend a low early-age test result.

7. The sampling procedure should also include measuring and recording slump and concrete temperature, and also cylindercube density on receipt at the laboratory. This is because such information is less expensive to obtain than the compressive strength, yet at least doubles the value we can extract from it. Entrained air tests are also useful, but this test is a little more expensive so it is not invariably justified. J.M. Shilstone (1987) has suggested that the fresh density of concrete may be a better quality indicator than slump. If taken it should certainly be combined with an air content determination, but it involves on site weighing equipment and it is not so simple to attain the required precision. Also it is not such a direct check on the relative water content of successive loads. It may be that hardened specimen density is sufficient providing that it is measured on receipt of the specimens at the laboratory (i.e., within 24 hours) and that it is immediately followed up by air testing when a significant density change is experienced. It may be that fresh density measurement is

mainly of use if rejection of trucks is contemplated, but this should be abnormal.

8. The test results should be analysed to detect, at the earliest possible time, any departure from the previously acceptable concrete properties. This can best be done by drawing cusum graphs of early-age and 28-day results, slump, temperature, cylinder density, 28-day pair difference, and early-age to 28-day strength gain.

   Such graphs are of substantial value not only in showing a strength downturn quickly and obviously but also in making it much easier to see whether the downturn is due to basic concrete quality, weather conditions, site abuse (excessive waiting time, water addition, etc.), or only the testing process.

9. It is very desirable to separate the functions of mix amendment and contractual acceptance. Mix amendment should take place based on early-age results and can be reversed without excessive cost having being incurred if found unnecessary a few days later. It can therefore be done on relatively slender evidence. Contractual acceptance is best regulated by a cash penalty or cash bonus based on a statistical analysis of at least thirty 28-day results.

   Physical rejection of hardened concrete, or even its further investigation by coring and so on, should be virtually unnecessary if these recommendations are followed. One very desirable result of a cash penalty/bonus specification is that it avoids any need to argue about a possible mix amendment based on slender evidence at an early age. The decision can happily be left to the supplier, as it is his penalty/bonus that is at risk rather than the structural integrity of the concrete.

The implementation of the aforementioned principles enables excellent control of concrete quality at very low sampling frequencies. The reduced volume of testing easily pays for the analysis, but much larger savings are made by the elimination of disputes, investigations, delays to program, rejections, and so forth. Day (1989) certainly did not advocate a greater expenditure on control by adding the cost of elaborate analysis to the cost of the present level of testing. The proposal was rather to minimise the total cost of a given degree of assurance of concrete of a given minimum quality. This cost includes the necessary minimum cost of the concrete, any extra costs imposed by restrictive specification requirements; the cost of testing; the cost of test result analysis; and any costs imposed by failures, including further investigation, partial demolition, legal costs, program delays, and wasted time in meetings.

A rapid check on water content of the fresh concrete using AASHTO T318 microwave procedure is another useful tool available to the concrete technologist.

7.3.4 Rounding results

It is extremely bad practice in any technical field to fail to recognise and take account of the inaccuracies inherent in test results. One aspect of this is to avoid expressing results to more significant figures than their accuracy justifies. In accordance with this various authorities require that certain test results be rounded. An example is the Australian NATA, which requires that compression test results be rounded to the nearest 0.5 MPa (= 75 psi) and densities to the nearest 20 kg/m³ (= approximately 1 lb/cu ft). Ken believes that this practice requires reconsideration.

Take compressive strength. Why should 0.5 MPa be selected? The answer is not that this is the order of accuracy, because different (competent) laboratories can easily differ by 2 MPa and average pair differences can exceed 1 MPa. Rather the answer is that in the days before computers were used, results were worked out from tables and 0.5 MPa steps gave about as large a table as was convenient. The tables would have been five times as large had 0.1 MPa been selected.

The important question is what use is to be made of the test result. Originally the answer was to accept it as totally accurate and reliable, and compare it to the specified strength. From this viewpoint it should certainly be taken as ±2 MPa and so labeled.

It is bad practice to round calculations before the very last step. The strength of the individual specimen used to be the last step, but now we have hopefully realised that this should no longer be the case. Action on compressive strength results should always be based on the analysis of groups of test results, effectively ignoring individual results. So it is the mean and standard deviation of a number of results that has significance. It would be better to use less rounded results, but it may not make a great deal of difference. However, when analysing (as we should) such items as within sample ranges (based on average pair differences) and 7- to 28-day strength growth, rounding to 0.5 MPa is obviously unsatisfactory.

It is proposed for compressive strength that it be expressed to 0.1 MPa and given the written qualification ±2 MPa where appropriate. This (apart from the ±2 MPa) will not consume any more paper and will marginally reduce the computer program.

For density, a similar situation exists. It is not so much the absolute density of a single specimen that should be of interest, but the range of densities of all specimens from a single sample of concrete (since this will reveal the competence of the specimen casting and enable its variation to be monitored). Detecting any change in the average density of concrete being produced, that is, of a group of samples, is the major reason for the test.

The proposal for density is that it be expressed as a four-digit integer, since again this takes marginally less computer effort and no more paper. The accuracy limits in the case of density may be much different for different organisations. For those to whom it matters, their control system will be providing a within-sample standard deviation. Density is unlike strength in that small variations in assessment of the same concrete by different laboratories are probably unimportant. Detection of change in average density or change in within-sample variation is probably what matters.

To accurately measure density, the authors strongly advocate weighing the test specimens in water and air, as the principle source of errors is the calculated volume, especially for cubes.

7.3.5 Cubes versus cylinders

The world is divided as to whether it is better to assess concrete strength by cube or cylinder specimens. The United Kingdom, much of Europe, the former USSR, and many ex-British colonies use cubes; the United States, France, and Australia use cylinders. The advantage of cubes is that they are smaller and do not require treatment (capping) prior to testing. The advantage of cylinders is that they are less dependent upon the quality and condition of the molds and that their density can be more readily and accurately established by weighing and measuring.

Both proponents naturally feel that the specimen with which they are familiar is preferable. The debate should be settled on the basis of which gives the most accurate (i.e., repeatable) result. This is best judged by the average pair difference achievable or the average range of three. Either of these can be converted into the within-sample (sometimes called within-test) standard deviation. In the case of pairs the average pair difference is divided by 1.13 to obtain the within sample σ. For the average range of sets of three, the divisor is 1.69.

Day received his initial concrete QC experience in the United Kingdom on cubes, and has owned and operated testing laboratories in Australia using mainly cylinders and in Singapore using mainly cubes. Both specimens are perfectly satisfactory and capable of very low pair differences if used carefully and cast in well-maintained molds. The problem is that the test specimens must be prepared in the field by relatively low-level technicians. The quality of training provided is crucial and is often inadequate. The really basic fault is often that the people training the technicians have inadequate knowledge, practical experience, or dedication to the task.

Capping used to be something of a problem with cylinders, although more of an initial than a continuing problem. It is still a problem in areas where cubes are the predominant specimens and the testing laboratories rarely test cylinders or cores. Once the proper equipment is obtained and the operator has gained experience, capping was never much of a problem. The capping referred to is the use of a molten sulfur mixture to achieve a smooth test surface on the end of the cylinder.

The essential items are

1. A heavy, accurately machined steel mold into which to pour the sulfur mixture
2. A guide along which to slide the cylinder to ensure the cap will be perpendicular
3. A thermostatically controlled melting pot in which to heat the sulfur mixture
4. A scoop holding an exactly suitable amount of the mixture to produce a cap

There are a number of difficulties to be overcome by the uninitiated:

1. Neat (undiluted) sulfur is not suitable because it shrinks too much and sets too quickly. A mixture with finely ground silica, fly ash, or other inert material should be used. Proportions are trial and error, depending on the particular sulfur and the particular filler. Some like to include a proportion of carbon black. Commercial blends are available.
2. The temperature of the mixture must be just right; too cool and it will not flow and set too quickly giving a thick cap, too hot and it goes rubbery and shrinks too much. Again it is trial and error.
3. The first cap is difficult because the mold is cold; later the mold gets too hot and causes delay waiting for setting.
4. The mold must be very lightly oiled between each use.
5. The cap must be thin, preferably less than 3 mm.
6. Especially for high strength concrete, a sulfur cap will not overcome a rough cylinder end. The cap will exhibit slight plastic flow under load and allow load concentration on high spots.
7. The hot sulfur emits fumes and requires at least an exhaust fan and preferably a fume hood.

All the above makes it quite clear why users of cubes are not tempted to turn to cylinders but has no bearing on the question of which is the more reliable test.

According to the U.S. Federal Highway Administration (FHWA-RD-97-030), traditional sulfur capping is suitable for concrete strength up to about 50 MPa. However, Lessard et al. (1993) found that a capping compound with a mini-cube strength of 55 to 62 MPa gave comparable results to grinding when used for testing concrete strength up to 120 MPa. However, the capping layer was less than 2 mm thick, which ensured adequate confinement of the capping compound. Under these conditions, the confined strength of the capping material is two to three times the unconfined strength. Lessard et al. suggested higher strength capping materials, and specialised preparation would only be required when testing strengths greater than 130 MPa.

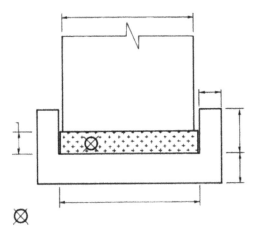

Figure 7.1 Rubber cap and restraining ring.

A significant improvement is that of rubber caps. The rubber cap system (consisting of a restraining cap and a rubber pad insert to accommodate irregularities) is a practical and cost-effective alternative to sulfur capping, which can be used by any laboratory. A suitable side clearance is essential since, under the high pressure, the rubber behaves almost like a fluid. If the clearance is too great, the neoprene will be extruded and will provide excessive side restraint. The mold is illustrated in Figure 7.1. A study by Carrasquillo and Carrasquillo (1988) showed that for concrete strengths up to 75 MPa, the use of neoprene inserts with steel restraining caps yielded average test results within 3% of cylinders using sulfur caps. However, the use of a neoprene capping system became a problem at higher strength.

A relatively recent development that could be very important is a new capping technique called the "sand box", although Day has heard no more of this development since including it in the previous edition of this book. The test was developed by Claude Bouley and Francois de Larrard, and was reported in *Concrete International* (Bouley and de Larrard, 1982). The "box" in question is a circular cup, very similar in appearance and function to the restraining ring used in the rubber cap test but deeper (30 mm). The rest of the apparatus is a positioning frame and guide similar to that used in sulfur capping, except that a small, air-driven vibrator is incorporated. The technique is to place a 10 mm layer of dry sand in the cup, position the cylinder in the frame, and vibrate so that the cylinder compacts the sand (20 seconds). The cylinder is then sealed into the cup by filling around the periphery with molten paraffin wax.

The test may initially look unattractive compared to sulfur or rubber caps, since it involves a capping process with molten material, a vibrator, and does not permit reuse of the mold before testing. However, it does not appear to involve as much manual dexterity as sulfur capping, avoids sulfur fumes, and permits immediate testing of a prepared specimen. It uses only sand and recyclable wax and so should be inexpensive in use. More important, it appears to give test results on very high strength concrete only slightly less reliable than the best achievable by end grinding and much better than even slightly substandard grinding. The trials have included successful use on extremely rough cylinder ends that would have had to be sawn off before any other technique could have been used.

The use of large aggregate concrete, except for special uses such as dams, is becoming rare. For high strength concrete, aggregate with a maximum size of more than 20 mm (¾ inch) is a disadvantage and for very high strengths a smaller size still, 10 to 14 mm (3/8 to ½ inch) gives better results. Therefore previously used specimen sizes of 150 mm (6 inch) cubes and 150 diameter × 300 mm long cylinders can be replaced by 100 mm cubes and 100 × 200 mm cylinders. Some researchers consider that the smaller specimens will give higher strength (up to about 5% higher) and greater variability. Others find that smaller cylinders give lower variability, but the differences are not sufficient to concern us unless they affect a comparison between different laboratories.

While considering such matters, reference must be made to the cube/cylinder ratio. A previous British Standard nominated this ratio as 1.25 for all circumstances, but this is not the authors' experience, which is that the ratio varies from over 1.35 to less than 1.05 as strength increases. A formula giving results in accordance with the authors' experience, but not claimed to be thoroughly established, is

Cube strength = Cylinder strength + 19/√ (cylinder strength) or

Cylinder strength = Cube strength – 20/√ (cube strength)

where cube and cylinder strengths are both in megapascal (MPa) or newton/square millimeter (N/mm^2).

Table 7.1 gives an alternative version that has greater official standing. The smaller cylinders, which weigh around 4 kg rather than 13 kg for the larger ones, are much easier to handle and cap.

With the advent of much higher compressive requirements, the preparation and type of capping becomes even more important. The best procedure is grinding the ends of the cylinder or at least the cast face of the cylinder if the base is suitably flat. Therefore it is the concrete alone that is tested without capping. Unfortunately this involves expensive equipment and may not be practical for field applications. However, for high performance concrete where a high level of quality control is required, we would recommend the use of ground cylinders rather than cubes or capped cylinders to minimise testing error.

Table 7.1 Cube and cylinder strength conversion

	Compressive strength at 28 days MPa (N/mm)	
Concrete grade	Cylinders (150 mm dia. × 300 mm)	Cubes (150 mm × 150 mm)
C 2/2.5	2	2.5
C 4/5	4	5
C 6/7.5	6	7.5
C 8/10	8	10
C 10/12.5	10	12.5
C 12/15	12	15
C 16/20	16	20
C 20/25	20	25
C 25/30	25	30
C 30/35	30	35
C 35/40	35	40
C 40/45	40	45
C 45/50	45	50
C 50/55	50	55

7.4 MATURITY/EQUIVALENT AGE CONCEPT

Concrete gains strength with age. It also gains strength more rapidly the higher the temperature. It is desirable to establish a relationship between strength, time, and temperature so that the strength of a particular concrete after any particular time and temperature cycle can be established from a knowledge of its strength after any other time and temperature cycle.

There have been two attempts to achieve this and both are detailed in ASTM C 1074. Although the two terms maturity and equivalent age are sometimes used in a qualitative way as interchangeable, they each have a precise meaning in numerical terms.

Maturity is the age of a particular concrete expressed as degree hours, that is, as the area under a temperature–time graph.

Equivalent age is the age at which a particular concrete would have developed its current strength if maintained at a nominated standard temperature.

Both of these definitions are incomplete in that the base temperature in the case of maturity, and the standard temperature and an "activation energy" in the case of equivalent age remain to be nominated.

The maturity (or TTF, time temperature function) concept was developed in the United Kingdom in the 1950s and is generally attributed to Saul (1951) or Nurse (1949). The base temperature should theoretically be that temperature at which concrete does not gain strength. This is often taken to be +10°F or –12°C. It is also often taken as 0°C for convenience, although concrete does gain strength at 0°C (but see Figure 11.2 and associated explanation for selecting a different value).

The equivalent age (EA) concept is older and more accurate, but also more complicated. The concept was not originated specifically for concrete but as a general concept for all chemical reactions. The general formula is attributed to Arrhenius. The concept was applied to concrete in the 1930s in USSR in the form of coefficients by which the length of time at each temperature should be multiplied to give equivalence.

The relationship is exponential and is given by the formula:

$$EA = \sum \left(te^{-Q(1/T_a - 1/T_s)} \right)$$

where

EA = Equivalent age (hours)
Q = activation energy divided by the gas constant
T_a = temperature (°K) for time interval t
T = Time (hours) spent at temperature
T_s = reference temperature (°K = °C + 273)

The reference temperature (T_s) is the standard curing temperature at which test specimens are kept. In many parts of the world it is 20°C (293K) in Australia it is 23°C in temperate zones and 27°C in tropical zones; it may be that 30°C would be appropriate in some tropical countries (if this is the average temperature of unheated curing tanks). The Q value can range from below 4000 to over 5000 depending on the characteristics of the particular cement. It is often taken as 4200.

A discussion of the relative merits of these two approaches follows, but it is important for the general reader not to get lost in the detail and worried about minor pitfalls, but to realise that the basic concept is very simple and enables powerful solutions to two problems:

1. Prediction of 28-day strength from an early-age test
2. Establishment of the strength of in situ concrete

Previously the first problem was approached by setting down a fixed accelerating (heated curing) regime and experimentally determining a correlation curve. The second problem used to be handled by setting a time, such as 7 or 14 days before some activity such as stripping, depropping, stressing, or lifting was permitted. Alternatively "field cured" specimens were used, assuming that cylinders cured alongside in situ concrete would have a similar maturity. This of course is very far from the truth. Almost any sort of rough application of any maturity approach is vastly superior to these "old-fashioned" solutions.

An initial approach to implementing the "new" concept was to construct a strength–maturity or strength–equivalent age curve experimentally in the laboratory. Having logged in situ temperatures, either the maturity or equivalent age could be determined at any time and the corresponding strength read from the graph. The weak aspect of this technique is the basic assumption that the in situ concrete is of identical strength to that previously used to construct the curve. There are two problems with this: One is that concrete is a variable material so that identical mixes are subject to a spread of results. The other is that there could (hopefully rarely) be a substantial problem with batching or quality of materials that would not be picked up by this approach.

The question of the accuracy of the two rival approaches (TTF and EA) arises. It seems to be generally conceded that the EA function is correct in that the effect of temperature is exponential rather than linear. However, opponents of using this approach point out that very large errors can result from an incorrect value for activation energy, whereas TTF is conservative. This is illustrated in Figure 7.2, which compares the two concepts and also the effects of varying the activation energy in EA and the datum temperature in TTF. The comparison is also affected by the standard curing temperature in use. In the illustrated cases these are 23°C and 40°C. This is the temperature of the water bath or fog room in which the specimens used to establish the standard maturity curve are kept. The assumption is that the true curve is between the two estimates of the Arrhenius curve and it can be seen that substantial

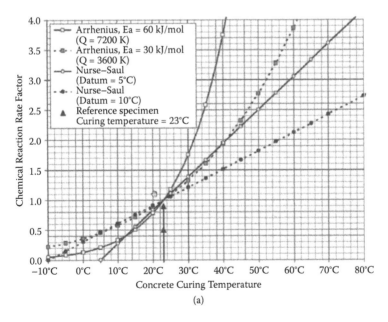

Figure 7.2 (a) Graphical comparison of maturity and equivalent age functions (23°C).

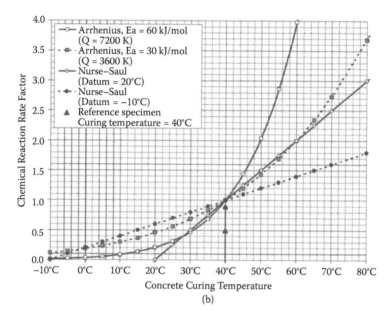

Figure 7.2 (Continued) (b) Graphical comparison of maturity and equivalent age functions (40°C). (Courtesy of Dr Steve Trost, Director, R&D for Strategic Solutions International, LLC.)

error could occur if the wrong one were chosen. It can also be seen that using the "correct" datum of −10°C for TTF is unconservative for temperatures below the standard curing temperature and overconservative for higher temperatures. Changing the datum to +5°C, while not theoretically correct, both avoids unconservative readings at low temperatures and reduces the overconservatism at higher temperatures. If the reference specimens are to be kept at any other temperature than 23°C, then Trost recommends that the TTF datum should be set at between 18°C and 20°C below the reference specimen temperature, specifically 5°C for 23°C and 20°C for a 40°C reference temperature. The reason for, and effects of, this can be seen by comparing graphs in Figure 7.2a and 7.2b. It is unlikely that anyone would use a reference temperature of 40°C (however, 27°C or even 30°C is normal in tropical countries) except possibly in the case of steam curing. However, it can be seen from Figure 7.2b that in the steam curing case, the TTF assumption, even with the 40°C reference temperature, would be likely to be underestimating true maturity by a factor of around 2 (with an average activation energy of, say, 4,200 and a steaming temperature around 80°C). This would be a substantial disadvantage in heating cost or cycle time and the only competitive alternative to using Day's EA system would be temperature matched curing (TMC). This may be a reasonable solution in a precasting factory but not otherwise.

However, TMC would not provide an early prediction as would Day's system and the TMC equipment would be more of a hindrance to production.

Day's approach to the concept was from the viewpoint of the other problem—the prediction of 28-day strength from early age tests. Guo (1989) suggested that using EA directly for this purpose did not work very well, but that a good prediction of 7-day strength was obtained. This is not surprising, given that Day has clearly established that predicting 28-day strength as a percentage of 7-day strength does not work very well either. Day's concept therefore was to use EA to predict 7-day strength and then, as in his normal control system, to predict 28-day strength by adding the average gain for 7 to 28 days (a figure already, and continuously and automatically updated, in his control system).

If the relationship between strength and EA is truly exponential, then a graph of strength against log(EA) will be a straight line, regardless of the activation energy. The slope of this line, the Q value in the EA formula, could be determined by entering any two results. In general, the average 7-day result is likely to be already in the system and being continuously updated. So testing a specimen at any early age, the slope of the graph between this point and the 7-day result can easily be determined. When the actual 7-day result from the same batch of concrete becomes available, the slope between these two results from the same batch can be substituted for the initial value and used to project the result from the next early age test to give a 7-day prediction and thereby a 28-day prediction. It was simple to write a program to automatically average all the Q values obtained in this way. The program also graphically displays all the slopes (Figure 7.3) so that it can be seen whether a consistent value is being obtained. Also it is very obvious if one or two results are clearly in error and these can be deleted. The system can also display a graph (cusum or direct) of the variation of the Q values so obtained, in the normal ConAd QC system, permitting a consideration of what factors might be influencing any observed changes in Q value. So the system provided an apparently foolproof means of applying the EA concept.

As often happens in concrete technology, things were not quite as simple and foolproof as appeared theoretically likely. Some inconsistencies were experienced and, in investigating them, many graphs of strength against log(EA) were drawn with multiple specimens from a single mix. It was found that, in all cases, the results formed not a single straight line but a broken straight line, that is, a straight line with a single change of direction somewhere along it. The change has subsequently been detected at anywhere between 2 and 7 days, and can occur in either direction (although almost always from an initially steeper slope to a later shallower slope). It is understood that the hardening of concrete is not a single chemical reaction and what this means is that two combinations of reactions with different activation energies are involved at different stages of the hardening process. Farro Radjy has used his "heat signature" technique to quantify

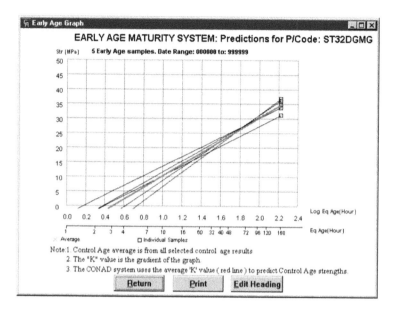

Figure 7.3 Automatic updating of K value (slope of strength versus log equivalent age graph).

the proportions of different chemical compounds present in a cement by their different rates of heat generation.

It is immaterial to our purposes what happens after 7 days, since this is covered by the addition of the average 7- to 28-day strength gain for the particular mix. However a change of slope prior to 7 days would mean that the slope of the line joining an early result and that at 7 days on the strength versus log (EA) graph would be affected by the particular early age. It was seen that the requirement for satisfactory operation is that the graph should be a straight line up to a predicted "control age" and that the amount to be subsequently added should be the average strength gain from the control age to 28 days. Therefore it is recommended that, for any new mix, a number of specimens are taken from a single batch of concrete, tested at a range of ages, and plotted on the strength versus log(EA) graph to determine the age at which the slope changes for that particular mix. This of course is quite different to the practice of using a predetermined strength versus maturity graph. There is no suggestion that subsequent mixes will have the same strength at the same EA, only that the rate of strength development will change at the same EA.

If the change in slope occurs at later than 7 days, then it will be convenient to continue to use 7 days as the control age. It is only when the change occurs earlier that the control age must be changed to get accurate results.

Having selected or established the control age, one specimen is always tested at the control age and at least one at some earlier age. The ConAd program automatically calculates the slope of the line joining these two points (we call this the K value) as the results are entered in the normal test result entry system (but having selected early age in the setup program). The program continuously averages all previous K values for the grade in question as each new result is entered.

Using the previous average K value, the control age strength is predicted as soon as the early result is entered. When the true control age result is later obtained, the true K value for that sample is evaluated and included in subsequent averages.

The system can display and print out the graphs and these can be used to establish the age at which any particular strength will be attained or the actual strength at any particular age. However, it is simpler and more accurate to use dummy results in the test result entry system. The nominated strength is entered and the age varied until the previous average control age (and 28-day) strength is predicted. Alternatively the nominated age is entered and the strength entry varied until the anticipated eventual strength is predicted.

An important use for the graphs included in the early age section is to check that the results do conform to a reasonable pattern. If Figure 7.4 shows some lines of distinctly different slope, then a problem exists. If no bias or particular period causing this can be found, then it may be due to testing error (the error can be either in the strength or the equivalent age

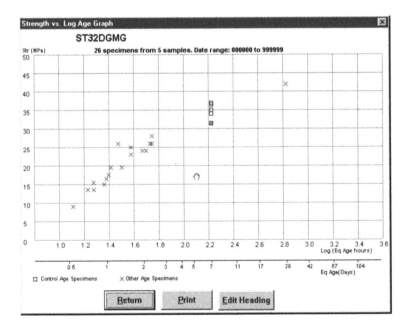

Figure 7.4 Strength versus log equivalent age graph.

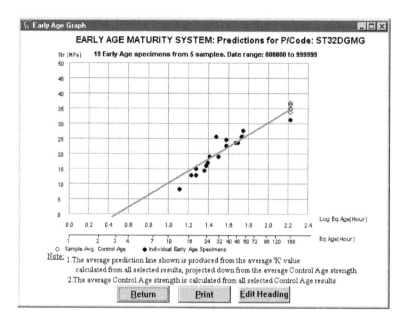

Figure 7.5 Early age specimen results. Strength (MPa) versus log equivalent age (hr).

of the early result). K values can be plotted in the normal QC system and it can be seen which results are abnormal. Such results can easily be excluded from the analysis by finding them in the table of all results (in Full View in the ConAd program) and so labeling them.

Figure 7.5 also clearly shows how much scatter of results there are and whether they are scattered about the correct line. For example the illustration chosen here does show some tendency to change slope at earlier than the 7 days selected as a control age.

Figure 7.3 is the kind obtained with multiple test ages and again shows a change at 2 days rather than the 7 days used. The error due to this assumption is quite small in this example but can be larger in other cases.

7.4.1 Limitations of the equivalent age concept

Concrete that has been heated

1. too early, or
2. too rapidly, or
3. too hot

will attain a lower 28-day strength than the same concrete cured at normal temperatures. The limiting values to avoid such problems differ for different cements, and especially for different combinations of pozzolan and cement. It does not follow that routines that involve a loss of 28-day strength should not be used; only that the loss should be understood and allowance made for it if necessary.

It can be anticipated that concrete containing a pozzolan or ground-granulated blast-furnace slag (GGBS) will withstand higher curing temperatures without loss of potential 28-day strength. Such concretes may show an increased 28-day strength through higher temperature curing.

Any particular curing regime for any particular concrete can be readily checked by comparing the strength versus logarithm of equivalent age curves for heated and normally cured test specimens. As a rough guide, a delay of 2 equivalent hours at 20°C, a rate of rise of 0.5°C per minute and a maximum temperature of about 70°C will usually avoid any significant loss of 28-day strength when using normal Portland cement.

Carino (1984) concluded that a parabolic relationship may be simpler to use and equally, or even more, accurate than the Arrhenius relationship. We have not experimented with such a relationship since it is easier to continue using the Arrhenius relationship now that it has been incorporated in a user-friendly computer program.

7.4.2 Temperature effects

It is necessary to protect concrete from freezing and thawing damage, and also from dehydration, until it has attained a critical strength beyond which further protection is less important. This has been recognised for many years and various national codes have laid down specified periods of protection. In some cases the protection period is varied according to ambient temperature, but much greater precision and flexibility is now feasible by defining the protection period in terms of measured equivalent age or of in situ strength determined from equivalent age.

Anecdotal evidence suggests that a similar situation may occur at high temperatures, although to a different extent. Thus a concrete specimen that is cast hot and stays hot until it attains substantial strength, or is heated and stays heated, may be less damaged than one that is cast hot and taken into an air-conditioned laboratory.

The possibility is that changing temperature may cause bond stresses at the paste–aggregate interface or microcracking in the paste or mortar fractions. It seems likely that such events would have greater significance for tensile and flexural strength, and for durability, than for compressive strength. A thermally caused reduction in compressive strength may be the tip of an iceberg in terms of total resulting damage.

Early exposure to high temperature without adequate protection often results in desiccation of the specimens. This can be a particular problem in higher quality concrete where hydration water lost prior to curing may not be fully replaced. The authors have observed many examples where this type of abuse of test specimens results in satisfactory early age strength but limited or no longer-term strength development. Most concrete standards have strict requirements for temperature, curing and the minimum age of demolding to help ensure that the test specimen provides a good estimate of the potential properties of the concrete. It would be a miracle if the concrete sampling as shown in Figure 7.6 with cubes exposed to high temperature with no curing and transported over site roads to the laboratory gave a true representation of the concrete properties. Unfortunately this is the norm in many areas and not the exception, often due to the misguided requirement in many specifications for sampling at the point of discharge which somehow takes precedent over the much more important requirements to protect the specimens.

The authors have worked on major projects with simultaneous concreting in several locations. The best way to control concrete in these situation is a centralised heated or air-conditioned laboratory to which the concrete trucks come for sampling before going to the delivery point. It provides better control anyway, particularly when more than one concrete type is being used. Concerns regarding possible water addition between sampling and placing or other effects can be overcome by collecting a bucket of concrete during discharge and taking it to the central laboratory to make specimens. The same procedure can be used to assess the effect of pumping and so on. The insistence by some engineers on casting specimens on climb forms and slip forms can pose significant safety issues. Indeed, the results from such tests often do not fulfill the primary goal of compliance testing of the delivered concrete.

Figure 7.6 Cubes taken near point of discharge on remote site.

7.4.3 Update on maturity/early age

Maturity monitoring has become more popular as the principle and economic benefits are more widely understood and accepted, and as instrumentation becomes more sophisticated and affordable. What is surprising is that the less sophisticated, degree–hour maturity concept (abbreviated to TTF or temperature-time function) is more frequently used than the more scientifically valid Arrhenius early age (EA) concept. This is apparently due to three factors:

1. TTF is easier to explain and much easier to calculate.
2. The determination of the activation energy, as set out in ASTM C1074 is an onerous process and substantial error can result if undetected changes in this occur.
3. Day's concept, described earlier, of a continually automatically updated constant in the log EA versus strength relationship (avoiding the need for prior calibration), while used enthusiastically on diverse projects in several countries by ConAd licensees for over a decade, has yet to be adopted (or possibly comprehended?) by anyone else.

In evaluating early-age strength it seems not to be generally understood/realised/allowed for that concrete is a variable material. Although concrete can be produced with a compressive strength standard deviation of 2 MPa (300 psi), unsophisticated producers may easily experience a figure of triple this or more. So the 28-day strength of a batch of concrete can vary by 1.28 (or 1.65 depending on country) × 6 = say 7 to 10 MPa or 1100 to 1400 psi. Generally the higher the strength of a concrete and the larger the percentage of that strength developed at a given early age, so early strength, at least as a percentage of average early strength, could be expected to vary at least as much as later strength. So an early age strength might be expected to vary by say ±2 MPa or 300 psi at a given maturity/equivalent age (even if no batching or other errors occur). This puts concern over the accuracy of equivalent age determination in a proper perspective.

There are now a number of instruments on offer that will log temperatures and even calculate maturity or EA within the concrete. This obviously avoids the problem of theft or damage of the recording device at the cost of it being sacrificial. Some such devices even incorporate radio transmission so that not even wire access is necessary. Again, this is an advance in convenience at an increased cost. In general the cost of physical equipment tends to reduce with time, while their efficiency and the value of convenience is perceived to increase. On this basis it is a reasonable assumption that such devices will continue to become more popular.

There are many factors to be considered in choosing equipment. Principle among these is confidence in the knowledge, ability, and good faith of the marketer and, as consequence, the acceptability of the results to supervising authorities. Unconservative assessment is one aspect of the risk and unreliability of the equipment is another. Either of the EA and TTF methods can be made to give satisfactory results by a knowledgeable operator using any of the available equipment. However, some equipment requires greater skill, care, and understanding than others and this can be involved/provided in different ways and at different stages. Decades ago, Day achieved satisfactory results by personally making and installing thermocouples and assessing results. Any faulty readings were recognised as such and discarded. Judgments on readiness for stressing and so on were made in a full knowledge of current test results and circumstances and past performance and with appropriate safety margins. This does not mean that the methods used would be satisfactory if applied by the average site worker or an inexperienced young engineer.

It is difficult to generalise on the economics of alternatives.

1. In most cases the savings made from the information gained far outweigh the cost of the testing. To this extent whatever it takes to satisfy authorities is worthwhile.
2. Also, in many cases the results reveal a substantial margin between the strength developed and that necessary for the purposes envisaged. In such cases a large margin can be allowed for inaccuracy.
3. The cost of personnel is often a major factor. They may be involved in preassembly, calibration, installation, reading, evaluating results, and equipment recovery. The level of skill and ability required varies significantly between different equipment and different work scenarios.
4. The number of probes installed may be influenced both by their perceived reliability and the consequences of an occasional failure.
5. The risk of damage or theft of external equipment will vary considerably between different working scenarios (e.g., site or precasting factory), and even different countries and locations.
6. The curing situation, varying from in situ slabs in winter to steam cured precast units, may be an overriding factor.

There is clearly a need for one or more kinds of certification, but this also may not be easy to arrange. One kind is the training and certification of operators by equipment providers. Another is the certification of equipment providers (as opposed to particular equipment). However, it is not clear who would be sufficiently competent and independent to provide such certification. It would be important not to introduce regulation that could rule out satisfactory solutions.

Chapter 10 gives details of the use of early age data in the ConAd QC and other programs. While these programs can display the graphs described earlier, it is not necessary to use them in the normal course of events, except for checking purposes. Entry of a strength and its associated EA in the normal QC program provides predictions of 7- and 28-day strengths and a method of predicting the strength at any desired EA or the EA at which any nominated strength will be attained. For steam curing situations, the user is able to nominate maximum and minimum estimates of the decline of temperature enabling the system to advise when steaming can be switched off to provide a specified strength at a nominated actual time.

7.5 FRESH CONCRETE TESTS/WORKABILITY

Fresh concrete can be tested for workability, air content, temperature, density, and moisture content, and analysed to give its approximate composition. As in most matters connected with concrete, it is again very important to have a clear idea of exactly what it is desired to achieve before deciding which tests are worthwhile and which are not.

7.5.1 Workability

A large number of tests for workability have been devised. The previous edition discussed the subject in great depth and relied heavily on a book by G.H. Tattersall (1991). The late Tattersall was a very important figure in the understanding of workability. Briefly, his major contribution was the realisation that no "single point" test could adequately quantify the workability of concrete. He established that concrete is not a "Newtonian fluid" in which displacement is proportional to the applied force but rather a "Bingham body" in which there is an initial resistance to displacement followed by displacement proportional to further applied force.

The term *rheology* is used to describe this more complex behaviour of fresh concrete under different conditions. This principle is now universally accepted, the initial resistance being known as the *yield stress* and the proportionality constant for subsequent displacement being known as the *plastic viscosity*. Since some concretes may have a lower yield stress but a higher plastic viscosity than others and vice versa, they will be assessed to have a different relative workability depending on the force applied during the test. This is particularly important since the slump test essentially only measures the yield stress, and compaction by vibrator is mainly dependent on the plastic viscosity. Tattersall proposed that it was necessary to conduct a two-point test at two different degrees of applied force to estimate both the yield stress and the plastic viscosity. This concept has given birth to a number of rheometers that measure the resistance to rotating paddles, cylinders, or discs in a reservoir of concrete at different speeds.

Since the third edition, there has been a trend toward moving rheological measurements from the laboratory to the site, often associated with the use of self-compacting concrete (SCC). Figure 7.7 shows the use of the

Figure 7.7 The ICAR rheometer being used for rheology measurement on Burj Khalifa project.

portable ICAR device on the Burj Khalifa project. This provided useful information on the effect of plastic viscosity on the friction factor of the different mixes being used as seen in Table 7.2 and Figure 7.8. Another very interesting observation was the change in rheological properties before and after pumping up to 600 metres. The slump flow reduced by up to 150 mm during pumping up to 600 metres. However, the plastic viscosity was reduced to approximately half the starting value and the dynamic yield stress was approximately doubled. This is believed to be temperature related due to the heating of the chilled 80 MPa concrete during pumping. It has important ramifications as the reduced plastic viscosity could make certain very workable concretes or SCC unstable after pumping. It is important to remember that the viscosity and yield stress values calculated by different rheometers differ and the parameter guidelines developed by one machine are not directly applicable to another machine.

The authors continue to consider that the real eventual answer to workability control must be a device fitted to every concrete delivery truck on any significant project. Such a device has been developed and was described in the second edition. Day observed it in technically quite satisfactory operation several years ago but it does not appear to have been a commercial success. However, Sensocrete is a similar system developed in Canada that uses an embedded probe and software to provide real-time assessment of

Table 7.2 Summary of concrete pumping monitoring

	Pumping			Concrete delivery pressure (bar)			Output (M³/h)	Friction factor(*10⁻⁶)		Plastic viscosity (Pa.s)
				Start point at near pump	End point horizontal line	Hydraulic pressure (bar)		Horizontal	Vertical	
Grade	Height	Level	Hydraulic pressure (bar)							
C80A	~350.8m	B2~L100	275~285bar	135~150bar	120~130bar	30~35	3.5~4.5	2.5~3.5	50~70
C80/14	354.3~456.0m	L101~L126	270~280bar	135~145bar	125~135bar	30~40	1.5~2.5	1.5~2.0	40~55
C50/20	~585.7m	B2~L156	240~260bar	120~130bar	115~125bar	20~30	1.5~2.0	0.8~1.2	30~40

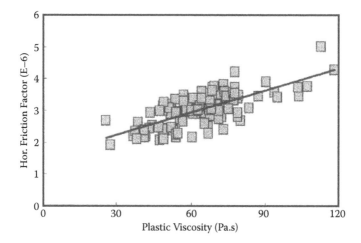

Figure 7.8 Relationship between plastic viscosity and friction factor.

rheology (http://www.sensocrete.com). Grace has also developed automated process control equipment for managing batches of ready-mixed concrete and reducing process variations (http://www.na.graceconstruction.com). The equipment reportedly accurately calculated the slump for a range of materials and mixture proportions and correctly adjusted the water content of individual batches to reach a target slump. At the ACI Convention in Dallas, Dr Denis Beaupre presented data on another truck measuring system (the IBB probe), which has been used in some production concrete and the results look encouraging. A comparison of the slump tests in accordance with the European and U.S. procedures showed the probe generated significantly lower coefficients of variation (2.8% versus 7.5% and 8.2%, respectively (Figure 7.9). This would be due to the number of readings possible for the probe. However, using such systems for compliance with slump requirements would appear to be missing the opportunity to get full benefit from rheological data. The fact that a number of suppliers are providing truck-mounted rheology monitoring systems is encouraging for future production quality control.

The problem with the slump test is that it is a very widely and firmly established test but is a poor measure of the relative workability of different mixes. It survives because of its simplicity and robustness and also because it is (when properly conducted) quite a good measure of the relative consistency (i.e., wetness) of successive deliveries of the same mix. With today's much more accurate batching and using the Day's mix suitability factor (MSF) we can have defined and controlled the other aspects of workability so that it may now be adequate to accept the slump test as defining consistency for the particular mix (especially if an "equivalent slump", adjusted for time delay and temperature, is used). What we must not do is to use

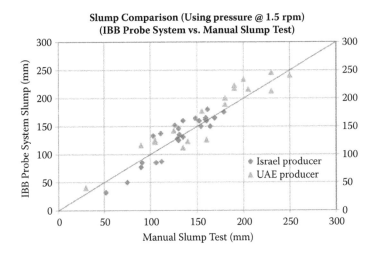

Figure 7.9 Comparison of measured and estimate slump from IBB probe.

slump in specifications on the assumption that it defines workability on an absolute scale. It may be acceptable for special purposes to specify slump limits in addition to precisely specifying the type of concrete required (such as special wear resisting floors) but generally workability (slump or otherwise) is the business of the concreter, not the specifier. The concreter should be permitted to strike his own balance between the higher cost of more workable concrete and the reduced cost of placing, always providing that such aspects as shrinkage, segregation, bleeding settlement, and so on are given adequate consideration.

Even the above half-hearted endorsement of the slump test does have its limits. Obviously it cannot be used for no-slump (or almost no-slump) concrete. Such concrete is likely to be used in precasting factories or roller compacted concrete (RCC) applications and alike. In such locations a V-B consistometer (AS1012.3, 1983) (in which essentially a workability test is performed in a cylindrical container and the time taken to re-form the slump cone into the cylindrical shape under standard vibration is measured) is likely to be convenient.

At the opposite end of the scale, flowing superplasticised/SCC/super workable concrete is becoming more popular. A flow table (DIN 1048) used to be the choice for accurate measurement of its workability. In this test it is the diameter of spread under a slight jolting motion that is measured. However with the higher fluidity now available, a simpler variant, the slump flow test (ASTM C1611) has taken over. The test also allows a subjective assessment of stability and a good indication of plastic viscosity with the T500 mm (20″) time.

The upper limit for which the slump test can be used is very dependent on the type of concrete. Harsh, gap-graded concrete (MSF of 20 or less; see Table 8.1) will fall apart on a slump test at slumps not much higher than 50 mm. On the other hand, continuously graded mixes of high sand content (MSF of 27 or more) will give a measurable and reasonably repeatable slump up to 200 mm or more.

The technique of carrying out a slump test is also important in obtaining a true reading and it should be realised that the slump itself is measured in different ways in the United States (to midpoint), United Kingdom (to highest point), and Australia (to average see 117).

What is important is not necessarily to stop using the slump test but to realise and allow for its limitations. For example a limiting slump value is often included in a job specification. With few exceptions, this is not the best way to achieve the specifier's objective. First, there should be an objective for the specification of anything, rather than it having been included in a previous specification and so mindlessly continued in the current document. The objectives may be to avoid high shrinkage, segregation, and bleeding, or to avoid an excessive water to cement (w/c) ratio leading to inadequate strength or durability. However, any of these faults can be encountered at almost any slump, however low, and avoided at any slump, however high. It is also easy to detect from a theoretical mix submission which mixes will be subject to one or the other of these problems. The contractor should therefore be permitted to submit his mix for approval at whatever slump he chooses, provided it is designed to accommodate his own slump limit without detriment. It is quite possible to produce fully flowing (250 mm slump or more) concrete having none of the potential faults noted and to produce almost all these faults in a 50 mm slump mix.

The rejection of a truckload of concrete on the basis of slump should also be approached in a reasonable manner. The slump test is both quite sensitive to small changes of water content and very easy to perform inaccurately. Certainly the truck driver should always be allowed to insist on the test being repeated. An extra 10 mm of slump probably involves about an extra 3 liters of water per cubic metre of concrete and may depress strength by about 1 MPa. The person charged with concrete acceptance should be kept continuously aware of the current strength margin of the mix in question and therefore of whether it is essential to reject slightly overslump concrete on strength grounds (and similarly for any shrinkage limit that may have been specified). It is more usual to find that a need to reject first arises on the grounds of wet properties or surface appearance. Slump variation will cause colour variation on a fair faced wall and slump in excess of that designed for can involve segregation, bleeding, delayed finishing, or floors of poor wear resistance.

Rejecting concrete for some petty reason such as a nominally noncompliant slump, temperature, and delivery time may result in an actual problem such as a cold joint or pump blockage as a result of the delay to replace the rejected truck. Unfortunately inexperienced supervisors are likely to reject nominal noncompliance without realising the damage that could be caused, particularly in high-temperature environments.

Although continuous perfection is impractical, a slump test will only be asymmetrical if it has been produced by an asymmetrical process. It is often possible to know where the slump operator has stood, how he has used his scoop, and how he has held his rod, all by looking at the resulting slumped concrete after the test. A failure to rotate the scoop will usually cause a higher coarse aggregate content opposite the point of discharge from the scoop. This will often cause the cone to lean toward the point of discharge on stripping. It is not easy to rod the foot of the cone opposite the operator if the rod is held in a dagger grip. To accomplish this the operator must project his elbow over the slump cone in order to rod each layer of the concrete parallel to the side of the cone around the entire circumference. An alternative is to use a rope grip, that is, to hold the rod as though pulling a rope.

The slump test is based on a standardised degree of semicompaction, unlike compression test specimens, which should be fully compacted whatever it takes. Therefore it is important that the correct number of strokes be used in the slump test while being only a required minimum in compacting compression specimens. It is also important that the rod have the correct end shape. A flat-ended rod (e.g., a piece of reinforcing bar) pushes coarse aggregate to the bottom and tends to leave a hole rather than compact. The British rod has a hemispherical end, which is a distinct improvement over a flat end. However. the Australian and American rods, which taper to half the original diameter before having a hemispherical end give greater compaction. It should also be realised that slump measurement is different in the United Kingdom, United States, and Australia. In the United Kingdom, measurement is to the highest point, in the United States to the point on the centerline of the original cone, and in Australia to the average of the original top surface. One may have personal preferences, but the important thing is to be consistent on a particular project and to be on the lookout for new operators who may have been trained by site engineers of different nationality.

A concept proposed by Day (1986) is that of an equivalent slump. As Bryant Mather (1987) has so firmly pointed out, slump loss is proportional to temperature and leads to the (strictly incorrect but workable) view that water requirement increases with temperature. Everyone realises that slump reduces with time. Putting the two effects together, it is clear that slump only has a real meaning if accompanied by a time and temperature reading. Day's proposal in the third edition was to combine the time and temperature into an equivalent age according to Arrhenius (see Section 7.4 on early-age strength for more detail). Thus an equivalent slump could be evaluated, being the slump that would be obtained had the concrete been kept for 30 minutes at a temperature of 20°C. It can be imagined that if compression speci-mens were stored at anywhere from 10°C to 30°C and tested at anywhere between 10 and 40 days, poor correlation would be obtained with w/c. This is what we are currently doing with slump tests (i.e., ignoring time and tem-perature effects). The development in admixtures and especially workability retainers that maintain workability for long periods without retardation does overcome the temperature effect to some extent.

It would be quite easy to arrange for a slump value to be converted into its equivalent value as it is entered into a computer, although less easy to arrange for this to be available during a field acceptance test. What becomes quite clear when these matters are considered is the absurdity of some rejection decisions currently taken in the field. A slump of say 150 mm taken 15 minutes after batching on a cold morning may indicate a lower water content, and therefore a stronger concrete, than a slump of 50 mm taken an hour after batching on a hot afternoon. Rules of thumb could be developed to provide some allowance approximately for this effect with at least more equity and realism than assuming that a slump is a slump and that's it.

With the above points considered, adequate attention given to correct sampling and remixing of the sample; correct bedding, cleaning and moist-ening of a rigid metal baseplate; and use of a square mouth scoop (because a round mouth scoop leaves mortar behind in the sampling tray) the slump test can give more reliable guidance than is often the case. Nevertheless one does encounter the occasional cheeky operator who asks what you would like the slump to be before carrying out the test. Suitably instructed, such persons are at least usually competent, since they obviously know what causes incorrect results

7.5.2 Assessing the workability of self-compacting concrete

Several special tests have been devised to measure the workability of SCC. These include the U box, L box, fill box, Orimet, and J ring in addition to rheometers, and are adequately described on the website http://www. efnarc.org (EFNARC being a European federation dedicated to specialist construction chemicals and concrete systems). These are essentially labora-tory tools to be used in devising SCC mixes and are too cumbersome to be likely to find site use except in major products.

The test likely to become the standard for site use is the slump flow test. This test uses the current standard slump cone but, instead of measuring the height of the cone, the diameter of spread is measured. The time for the out-ward flow to reach a diameter of 500 mm, (20"), known as the T500 time, is desirably also recorded. A further variant is to surround the slump cone by a steel ring of 300 mm diameter with evenly spaced "feet" of vertical 100 mm steel bars known as a J ring. The diameter and spacing of the feet can be varied according to the congestion of the reinforcement in the section to be cast. Some J rings are invertible with different spacing of feet according to orientation. Apart from a visual observation of the flow through the J ring, the depth of concrete inside and outside the ring can be measured.

For self-compacting concrete, a flow diameter of at least 550 mm is required, with a T500 time of 2 to 7 seconds. Visual observation of the edge of the spreading concrete is important. The concrete should appear to roll out with a blunt edge and no toe of fluid paste (which would indicate bleeding) advancing in front of it. Coarse aggregate must be present right up to the edge and evenly spread over the area of concrete. There should be no concentration of coarse aggregate in the center of the spread (which would indicate segregation).

Interestingly, the same diameter of spread is obtained whether the slump cone is used in its normal orientation or inverted (Procedure A or B in ASTM C1611, respectively). Although both alternatives currently have their advocates, it is clearly the inverted option that will survive long term for the following reasons:

1. The fluid concrete exerts a pressure on the sides of the slump cone mold. In the normal orientation this pressure has an upward component and, especially since the fluid contents leak very easily, the operator has to concentrate on maintaining downward pressure on the mold while filling. In contrast, in the inverted position the fluid pressure has a substantial downward component and can even be filled without being held in position (once partly filled).
2. In the inverted position the large open end is obviously easier to fill without spilling.
3. When using a J ring, the feet of the slump cone are a problem in the normal orientation.
4. Two operators are often used to obtain a T500 time, but it is possible to juggle a stopwatch when using the inverted position.
5. In the inverted position the T500 time is a little longer and so a little more tolerant of inaccuracy in timing.
6. Any tendency to segregation in the form of a concentration of stone in the center of the spread will be exaggerated by use of the inverted position.

So in summary, the inverted position is easier to use and is a slightly more severe (and therefore better) test.

7.5.3 Segregation resistance

According to EFNARC, SCC is defined as "concrete that is able to flow and consolidate under its own weight, completely fill the formwork even in the presence of dense reinforcement, whilst maintaining homogeneity and without the need for any additional compaction". Whereas most tests focus on the flowability or passability of SCC, the key consideration and distinguishing feature of SCC must be its segregation resistance and maintaining homogeneity. The most commonly used procedure is the visual stability index on the concrete perimeter after the slump flow test (ASTM C1611). The qualitative nature of the assessment makes it dependent on the experience of the tester. Another problem is that any effect can be masked by the presence of liquid water on the surface of the baseplate. The column segregation test (ASTM C 1610) is not suitable as a compliance or field test and therefore only appropriate for research.

The 5-minute V-funnel test is a useful method where a more than 3-second increase in flow time suggests the SCC does not have sufficient segregation resistance. The V-funnel test is described in Annex B2 of the EFNARC guidelines on SCC. The GTM screen stability test weighs the amount of mortar passing through a 5 mm sieve. This quick test is suitable for field use. ASTM C1712 "Rapid Assessment of Static Segregation Resistance of Self-Consolidating Concrete Using Penetration Test" is a practical field procedure with a suitable guideline for interpretation. Another important thing to remember is that segregation resistance can change over time when the effect of viscosity modifying admixtures wears off. Testing segregation resistance over time should be conducted to confirm that this is not a problem.

7.5.4 Compacting factor

The compacting factor test achieved a degree of success in the United Kingdom at replacing the slump test but is virtually unused commercially elsewhere and must now be regarded as historical. It is a device using two hoppers mounted above each other in a frame, with the lower hopper discharging into a standard cylinder mold. The concept is that the first hopper fills the second in a standard manner and the drop from the second hopper into the cylinder mold subjects the concrete to a standardised compactive effort. The result is expressed as a proportion of full compaction achieved by dividing the weight of concrete in the mold by the weight of a fully compacted cylinder.

The test is a little more accurately repeatable and is a more absolute basis of comparison between the relative workabilities of different concrete mixes than the slump test. However, the test is not greatly superior to the slump test in quantifying variations in water content of successive deliveries of the same mix, and since it is less widely used and involves more cumbersome and expensive equipment, it does not seem likely to survive. It may be reasonable to assume that if anything more elaborate than a slump test is desired, a portable rheometer is the way to go.

If a rheometer is not available, it is again emphasised that slump (or slump flow) plus an MSF (i.e., relative sandiness) and adjusted for time after batching and possibly concrete temperature, is a more meaningful measure of workability than slump alone.

7.5.5 Air content

Entrained air is generally used for two different purposes: (1) to improve resistance to freezing and thawing, and (2) to improve workability and inhibit bleeding. It may also be used for reducing density, especially when using lightweight aggregate.

For the freeze–thaw application a higher percentage (5% to 8%) is required than is normally used for workability improvement and bleeding inhibition (3% to 4%). At the higher percentage, entrained air costs money in the form of needing a lower w/cm ratio or higher cementitious content for a given strength and workability. At the lower percentage, and at concrete strengths of 30 MPa (4350 psi) and below, the water reduction enabled by the air entrainment may fully offset the weakening effect at a given w/c ratio. The water reduction may be of the order of 10% and the strength loss at a given w/c ratio about 5% per 1% of air entrained. It should not be forgotten that non-air-entrained concrete is likely to contain 1% to 2% of voids so that the extent of the extra weakening may be only 5% to 10%.

It should not be forgotten that frost resistance depends upon bubble spacing, whereas strength reduction is proportional to total air volume, so that it is highly desirable that bubble size is as small as possible.

It is obviously necessary to specify the required air content where this is 5% or more, since otherwise it would be omitted on economic grounds by the concrete producer. It would also be reasonable to regularly test the air content in this case.

Where the air is not required for freeze–thaw durability, it may be unnecessary to specify it. Partly because it may be provided in any case and partly because fly ash, with particles similar in size and shape to entrained air, has a similar effect (although a smaller water reduction). The amount of entrained air can be deduced reasonably accurately from the fresh density or hardened density of the test specimens (cube or cylinder). When this density indicates that the air content may have changed, it may be desirable to immediately institute air content testing until the reason for the changed density is established.

An air void analyser (AVA) provides information on specific surface and spacing factor of fresh, air entrained concrete within 25 minutes enabling compliance testing before placing the concrete.

7.5.6 Density

Some concrete controllers like to carry out regular fresh density testing. It is certainly true that there is often a good correlation between strength and density for a particular mix. However, as noted earlier, the density of hardened test specimens on receipt at the laboratory may be an adequate substitute for routine control purposes. Where the purpose of the density test is to settle a dispute on the yield of the mix (i.e., whether a nominal cubic metre is in fact a full cubic metre) it is certainly necessary to carry out a very formal, fresh density check. In any case it is desirable to carry out such a check initially or very occasionally to verify or modify the assumption that it is adequately represented by the hardened specimen density. In such a test it is very important not to omit the use of a glass top plate since, however carefully it is done, striking off level is never accurate enough (usually the measured density is too high without a plate, but it can be too low).

When such arguments get to very fine tolerances, the question arises as to whether the concrete supplier must provide a full cubic metre of hardened concrete. Obviously the purchaser is entitled to fully compact the concrete as regard entrapped air, but is he entitled to vibrate out some of the entrained air? Also, if the concrete displays bleeding settlement, is it the volume before or after this that counts? These differences are quite small, but in a situation where a great deal of concrete is placed with low labor and formwork costs (e.g., thick, unreinforced airport paving) they can constitute a substantial proportion of the profit margin. There is no correct answer to the foregoing questions, they are subject to negotiation, but it is as well to realise the situation if negotiating.

The correlation between strength and density arises because air and water are the two lightest ingredients of concrete and cement is (almost always) the heaviest ingredient. The only other factor likely to have influence is the specific gravity of the coarse aggregate. In lightweight concrete the moisture content of the coarse aggregate may also be a significant factor.

It is also good practice to accurately measure the hardened density of the cubes or cylinders. Whenever possible, this is best done by measuring mass in water and in air to eliminate errors in volume calculation.

7.5.7 Temperature

The cost of measuring the temperature of concrete at the time of casting test specimens is negligible, so it should always be done. The availability of accurate infrared temperature without even touching the concrete means that premix trucks can be assessed without the inconvenience of sampling the concrete. There is often a good correlation between temperature and strength (higher temperature, lower strength) arising mainly from the increase in water requirement at higher temperatures. However, it is possible that early-age strength will increase with increasing supply temperature, the additional maturity being sufficient to more than offset any increased water requirement. This is more likely to occur with say a 3-day test than a 7-day test and in cold climate countries rather than hot ones.

The temperature of the test specimens prior to demolding can have a significant effect, particularly on early-age strength. This is the reason that a properly controlled and managed testing facility on site is so important.

7.5.8 Moisture content

Earlier moisture probes for real-time assessment of moisture content of aggregate suffered from problems with clumping of aggregate around the probe. However, recent microwave-based probes with suitable calibration do appear to provide a good estimate of moisture content in aggregates during production.

In the third edition, Day suggested that with the low cost and ready availability of microwave ovens there should be an increasing use of measuring moisture content by drying a sample of wet concrete taken back to the laboratory. AASHTO T318 "Standard Method of Test for Water Content of Freshly Mixed Concrete Using Microwave Oven Drying" is a useful test for monitoring the water content of the fresh concrete. The procedure takes about 15 minutes and the single operator within-laboratory standard deviation has been found to be 1.6 kg/m^3. This procedure is particularly useful in controlling questionable premix suppliers. The Port of New York and New Jersey uses this procedure as a quality control procedure on its projects. Although the largest source of error should be in a nonrepresentative ratio of mortar to coarse aggregate in the sample, the AASHTO procedure seems to provide reasonable repeatability without sieving.

7.5.9 Wet analysis

The UK RAM (rapid analysis machine) is an apparatus designed by CACA to split a sample of fresh concrete into its constituent parts. It is well known but apparently little used outside the United Kingdom. According to Neville (2011), the RAM has not proved successful, and the ASTM test methods for the fresh cement and water content determination have been withdrawn. Clearly the increased complexity of the cementitious component in concrete due to the use of binary and ternary blends makes rapid wet analysis based on particle size or even chemistry difficult.

As regards the relative proportions of the dry ingredients, most plants and projects are served by a computerised batching plant that can provide a hard copy computer record of the batch weights, which should be able to settle any question of deliberate deception. Focusing on performance requirements rather than the specific ingredients seems the best way forward. The problems with specifying minimum cementitious contents are discussed in Chapter 6 and therefore, in our opinion, the quantity of cementitious material should be the concern of the premix company. The relative proportion of different cementitious materials can be more complicated as it may affect long-term performance. For example, in the Middle East, supplementary cementing materials are imported and generally more expensive than Portland cement. They are often specified at minimum replacement levels to reduce temperature rise or improve durability. Variation in the proportion of these materials can result in significant detrimental effects on the concrete, and any procedure that can detect such variation as soon as possible would be very helpful.

The mutual suspicion that often exists between specifiers and suppliers is counterproductive to achieving optimum performance and reducing variability to the benefit of all parties. Certainly, restrictive specifications that do not allow the premix supplier to adjust the mix in response to variations in materials or mix registers that fix mix proportions do not encourage transparency when it comes to information of actual batch weights.

Because of the large volumes involved in producing concrete and the errors associated with the analysis, the authors are not convinced that attempts to confirm the correct mix proportions by wet analysis techniques should be the focus of quality control except for critical parameters such as free water and air content described earlier.

7.6 TEST PROCEDURES FOR ASSESSING DURABILITY

We will focus on chloride-induced corrosion, which is the predominant reason for deterioration of reinforced concrete. The subject of test procedures to assess durability in severe environments is a veritable minefield where only the brave (or the foolish) dare to tread. However, a fundamental change in the way specifications are written seems necessary. The authors believe effective and user-friendly performance specification is the answer. The object of appropriate performance tests should be to provide an acceptable probability of achieving the specified service life. However, there are a number of problems with establishing an agreed-upon test procedure and performance criterion upon which to base such a specification. First, there are distinct transport mechanisms: sorption of water containing chlorides, permeation of the chloride solution, and the diffusion of free chloride ions. These may act singly, simultaneously or in series depending on the exposure condition and the moisture content of the concrete. For example, concrete in the lower splash zone when impacted by a wave will initially be exposed to permeation of seawater under some pressure. Simultaneously and then continuing for some time after permeation, sorption will occur into the unsaturated voids. As surface chlorides accumulate, diffusion will then draw ions deeper into the concrete provided there is sufficient moisture for continuous liquid pathways. Careful consideration of the ways in which chlorides could penetrate the concrete cover in different parts of a structure is a valuable starting point in helping solve potential durability problems. Secondly, commercial interests often obscure the relative importance of different transport properties. Manufacturers marketing products to enhance concrete durability will emphasise the importance of the parameter primarily influenced by their material. Finally, effective durability enhancement materials or systems that change the electrochemistry of the corrosion process, such as inhibitors, may be selected against by specifications focusing only on transport properties.

7.6.1 Compressive strength

Twenty-five years ago, Neville (1987) considered the use of concrete strength as a basis of its acceptance as a culprit in durability problems. He stated, "My submission is that we should have decreased our concern with strength long since and we should have concentrated on developing practical criteria for durable concrete which could be used in specification". Compressive strength on standard cubes or cylinders is routinely tested for compliance on most projects. Thus it is not surprising that designers have tended to use compressive strength as an indirect indicator of durability in the belief (or hope) that strong concrete will be durable concrete. Indeed, many codes set minimum compressive strength requirements for severe environments. AS 3600-2009 requires a minimum compressive strength of 50 MPa for tidal or splash zones. It also recommends a 65 mm cover and 7 days continuous curing. This minimum compressive strength requirement was developed to ensure the required sorptivity based on the comprehensive work done by Ho and Lewis (1984, 1988). They conducted an extensive program on the influence of a wide range of factors on the sorptivity or capillary absorption of concrete.

We agree that sorptivity can be an important factor in the durability of concrete in such environments. However the conclusions regarding the relationship between strength and sorptivity (Ho and Lewis, 1988) may have been influenced by the sample preparation procedure where the concrete specimens were air dried for 21 days from their saturated condition in the laboratory at 23°C and 50% RH prior to testing. As has been pointed out by Dolch and Lovell (1987), the processes of both drying and wetting are influenced by water to cement ratio and therefore strength. Thus higher strength concretes would dry more slowly. As the moisture content of the specimens at the start of the test will strongly influence their sorptivity (Concrete Society, 1988), the sample preparation used in Ho and Lewis' programs would tend to favor higher strength mixes.

7.6.2 Sorptivity

One popular sorptivity test is the initial surface absorption test or ISAT as prescribed in BS:1881:Part 5:1984, which has been withdrawn. Although the procedure and performance criteria were developed on oven dried specimens, the standard procedure allows testing after only 48 hours ambient drying. The limited drying of 150 mm cube specimens means that compliance testing based on ISAT results in very low values. In the Middle East, ISAT is used as one of four so-called durability compliance tests. The commonly used performance criterion is 0.15 ml/m²/s at 10 minutes (60% of the value recommended in Concrete Society TR31). On one project, over 150 separate ISAT compliance results taken from three different grades of concrete never exceeded 0.03 ml/m²/s or 20% of the performance criterion. Any test procedure with such limited discrimination is of no benefit in performance specification, as it will not highlight changes in performance.

For the New South Wales (NSW) road transport (RJA, now Roads Maintenance Service, RMS) authority, the sorptivity test procedure involves drying of 100 mm × 100 mm × 350 mm concrete beams at 23°C and 50% RH for 35 days if the concrete will be exposed to the tidal/splash zone. In 2011, the sorptivity test was effectively downgraded to only a curing effectiveness test. The performance criterion was 8 mm water penetration after 24 hours in contact with water. The Cement, Concrete & Aggregates Australia publication on "Chloride Resistance of Concrete" dated June 2009 states that "RTA sorptivity of 1 mm is equivalent to 0.026 mm/min$^{1/2}$, and RTA (RMS) sorptivity of 3.8 mm is equivalent to 0.1 mm/min$^{1/2}$". However, the test measures the depth of water penetration not the volume as other typical sorptivity tests such as ASTM C1585, Capillary Index, and the TE Sorption procedures. When converting from the RTA sorptivity one needs to consider the porosity of the concrete to convert from penetration depth to penetration volume.

The measured sorptivity is strongly influenced by the degree of drying of the test specimen prior to contact with water. The RTA (RMS) T362 procedure dries at 50% RH for 35 days. The ASTM C1585 procedure involves drying specimens for 3 days at 50°C and 80% RH and then sealing for 15 days. The 30 minute absorption, Capillary Index, and TE Sorption tests oven dry at 105°C. Each test procedure will give different relative sorptivity values with lower sorptivity for less intense drying. The justification for longer drying conditioning is that it is more representative of ambient conditions. However, increasing the time and complexity of a compliance test profoundly reduces both the number of test results obtained as well as the ability to respond to variation and noncompliance.

The time taken to complete the procedure due to the prolonged drying coupled with the approved concrete mix register has meant that RTA (RMS) sorptivity test was not used as an ongoing compliance test. Therefore statistical assessment of the variability of the test is difficult. As reported in

Concrete Society Technical Report 31, a study on the variability 110 separate 30-minute absorption tests on a single grade of concrete found an average of 2.2% with a standard deviation of 0.35% equivalent to a coefficient of variation of 15.9%. This was far lower than for the other penetrability tests.

Summers (2004) used the Capillary Index test for compliance testing of a major project in Bahrain. The average value for the 557 results was 4.4 × 10^{-4} vol/vol/s^{½} and variability is shown in Figure 7.10.

CSIRO (1998) found that the Road Traffic Authority (NSW) sorptivity was a reasonably good indicator of chloride diffusion coefficient after one-year exposure (Khatri et al., 1998). However, in the recently released version of B80, the sorptivity test is used to assess curing only, not durability. Chloride diffusion or migration coefficients are the current performance criteria. This change was not due to any durability problems associated with concrete conforming with the sorptivity requirement but due to the belief that chloride diffusion was the primary transport property of interest.

Another popular "sorptivity" test is the 30-minute absorption test (BS 1881:Part 122:1983). Measuring water absorption into concrete after a short period of immersion is effectively a sorptivity test, as it gives an indication of the rate of absorption before full saturation can occur. The test procedure is simple and relatively quick to perform (as the specimens are oven dried at 105°C) requiring no specialised equipment. One of the authors promoted this test for performance specification some 30 years ago and advocated testing at an earlier age to enable quicker response to variation. The test is used as one of the four so-called durability compliance tests in the Middle East. Unlike ISAT, the 30-minute absorption test does provide useful results.

Malier and Regourd (1995) established an accelerated wetting and drying test for chloride penetration. The procedure involves drying at 40°C

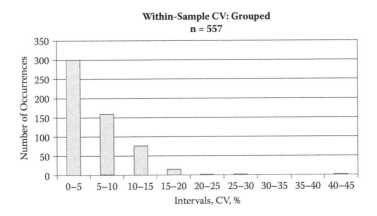

Figure 7.10 Variability of the capillary index test. (From Summers, C. R., *Concrete Tech. Today*, 3, 1, 22–29, 2004.)

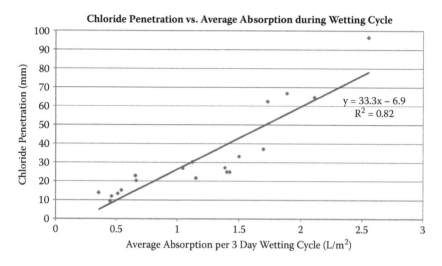

Figure 7.11 Relationship between chloride penetration and absorption. (From Aldred, J.M., Chloride and Water Movement in Concrete with and without Admixtures, M.Eng. thesis, National University of Singapore, Singapore, 1999.)

for 4 days and wetting with chloride solution for 3 days. Figure 7.11 shows the correlation between chloride penetration and the average absorption during the 3-day wetting cycle after 8 cycles (Aldred, 1999). Obviously, under conditions of cyclic wetting and drying, sorptivity is a dominant transport mechanism.

This is highlighted by cores taken from concrete with and without a hydrophobic poreblocking ingredient after more than 15 years exposure to daily wash-down with hypersaline bore water. The control concrete exhibited aggregate exposure and decalcification due to magnesium sulfate attack as well as significant chloride and sulfate penetration. The hydrophobic concrete exhibited no surface deterioration as well as limited chloride and sulfate penetration. The key difference between these concretes was sorptivity. The 30-minute absorption of the concretes with and without the hydrophobic admixture were 0.4% and 2.3%, respectively. The sorptivity values according to ASTM C1585 were 9.7 and 15.8×10^{-4} mm/s$^{0.5}$. The chloride activation depth (0.06%) in the hydrophobic concrete was approximately 15 mm compared to greater than 70 mm for the control. Therefore the 30-minute absorption provided a better indication of relative field performance in the field than the ASTM procedure where the specimens were tested at much higher internal relative humidity. The difficulty is determining an appropriate performance limit that would ensure acceptable durability without reference to specific proprietary products.

Sorptivity also appears to be a good indicator of resistance to physical salt attack, which is not surprising considering the mechanism of salt accumulation in a partially immersed condition. A 30-minute absorption limit of approximately 1.2% was found to provide good long-term performance, over 40 years.

7.6.3 Volume permeable voids (vpv) or porosity

Porosity tests achieve virtually full saturation of pores and microcracks by the completion of the test. Such porosity tests would include AS 1012.21, ASTM C642, RILEM CPC No. 11.1, and CPC No. 11.3.

The Australian and American standards achieve virtual saturation by using relatively small specimens that are placed in boiling water for considerable periods. RILEM CPC No. 11.1 involves soaking the specimen until constant weight gain. In RILEM CPC No. 11.3, the oven-dried specimen is placed in a vacuum followed by immersion in water.

Vicroads in Australia requires maximum VPV values for all grades of concrete. The range of maximum values for rodded cylinders varies from 12% for 55 MPa concrete with a maximum w/cm of 0.36 to 15% for 32 MPa concrete with a maximum w/cm of 0.5. The test method has been used for performance specification in Victoria for 20 years (Andrews-Phaedonos, 2012). Although advocates of this procedure point to the low coefficient of variation associated with the test (2.4% according to Whiting, 1988), one reason for the low variability is the very limited discrimination within the test procedure.

Comparing the data from the repeatability figures in Andrews-Phaedonos (2012), the variation in average 7 day VPV for Mix 4 (VR400/40) is 13.7% to 12.6% or a difference of 1.1%. The difference in average VPV for the same mix from 7 days to 28 days and 90 days water curing is 0.9% and 1.2%. According to the data presented, the potential difference between two mixes with the same proportions under laboratory conditions after 7 days curing is of the same order as the difference due to an additional 83 days water curing. The mix design and chloride diffusion coefficients at various ages are not presented. If the mix contained up to 25% fly ash or 40% GGBS in accordance with Vicroads Section 610.07 f, the chloride diffusivity may have reduced by up to 65% between 7 and 90 days whereas the VPV reduced by up to 10%.

Whiting (1988) compared different penetrability parameters with the 90-day ponding test (AASHTO T259) where chloride penetration is due to the combined effect of absorption, wick action, and chloride diffusion. Six concrete mixes with w/cm ratios ranging from 0.26 to 0.75 were tested and cured for 1 or 7 days before testing. Five of the mixes were pure ordinary Portland cement (OPC) mixes and one contained 11.7% silica fume. The correlation coefficient between the total chloride penetration into the concrete (2–40 mm) and VPV% was 0.90 as seen in Figure 7.12. Because Whiting did not measure the chloride profiles, it is not possible to estimate the chloride diffusion coefficients of the concrete mixes and the correlation with the measured VPV. The tests were also confined to concrete with

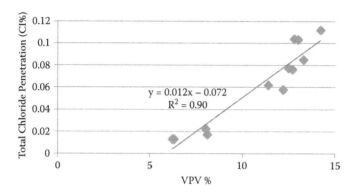

Figure 7.12 The relationship between chloride penetration (% by mass 2–40 mm) and VPV%. (From Whiting D., Permeability of Selected Concretes, ACI Special Publication, SP 108–11, 195–222, 1988.)

limited curing and not containing cementitious materials with significant binding capacity which would reduce the long-term reduction in chloride penetration as discussed earlier.

Sherman et al. (1996) tested concrete with w/cm ratios of nominally 0.32, 0.37, and 0.46 with and without silica fume. The concretes were exposed to different curing regimes—water, burlap, or heat curing—before testing. Plotting the chloride diffusion coefficients against VPV% showed no clear relationship (Figure 7.13). An interesting result from this research was the effect of the curing regime on VPV. Water curing compared to burlap, for 7 days reduced the average VPV by 8.4% as would be expected. However, concrete exposed to a heat curing regime that involved 7.5 hours at 63°C and storage at 50% RH after 24 hours reduced the VPV by an average of 25% compared to water curing for 7 days (see Figure 7.14). Elevated temperature curing is typically associated with reduced long-term performance of concrete and therefore a lower VPV demonstrates the limitations of using indirect indicators. This unusual temperature effect may be one of the reasons for the much lower maximum VPV requirement of 7% for severe exposure in the Middle East (Summers, 2004) compared to a maximum of 11% for Vicroads. However, other requirements such as a maximum binder content of 400 kg/m³ rather than a minimum binder content of 470 kg/m³ and a ternary blend should help improve durability and reduce VPV.

In our opinion, porosity tests provide additional information on the penetrability of concrete. Variations in measured porosity in a particular concrete mix will help detect a change in quality. The concerns with the porosity tests for performance specification are that they have limited discrimination, they are affected by concrete temperature, and they can be poorly correlated to key transport properties such as chloride diffusion and

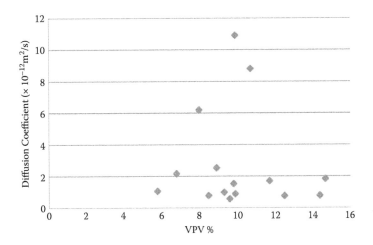

Figure 7.13 Comparison of chloride diffusion coefficient with VPV. (From Sherman, M. R. et al., *PCI Journal*, 75–85, July–August 1996.)

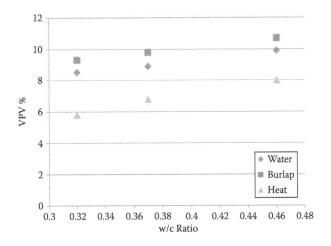

Figure 7.14 The effect of curing on VPV. (From Sherman, M. R. et al., *PCI Journal*, 75–85, July–August 1996.)

sorptivity. It is not the total porosity per se that determines durability but the rate of water and salt ingress. It is possible to have hydrophobic, lightweight, or high SCM replacement concretes with excellent durability but with a relatively high total porosity. Accordingly, the porosity tests are not particularly appropriate for assessing durability in severe environments but can be used as a quality control tool. The more information the better, within reason.

7.6.4 Water permeability

Water permeability is often used in a generic sense for all forms of water movement. However, the American Concrete Institute (ACI) defines it as "the rate of discharge of water under laminar flow conditions through a unit panel cross-sectional area of a porous medium under a unit hydraulic gradient and standard temperature conditions, usually 20°C". Therefore, water permeability necessarily involves the hydraulic pressure through saturated concrete.

Deterioration of concrete is most prevalent where concrete is subjected to wetting and drying or partially submerged. Under these conditions, the concrete is neither subjected to a hydraulic gradient nor saturated and thus permeability cannot be the driving force for water movement or salt accumulation. Even in tunnels and other structures exposed to hydrostatic pressure the effect of evaporation from the surface appears the dominant factor (Aldred, 2008). Aldred et al. (2001) calculated that water flux through OPC concrete with w/c 0.4 due to water permeability alone would be an order of magnitude less than that due to wick action (i.e., flow due to water on one face and air on the other without pressure) for a typical retaining wall of 300 mm thickness exposed to 20 metre hydrostatic pressure. Therefore, water transport through uncracked concrete under most practical situations is not dominated by the external hydrostatic pressure. Accordingly a water permeability coefficient as measured by the pressure differential tests would not appear the dominant driving mechanism of water or salt penetration in most severe environments.

Vuorinen (1985) found that oven drying and resaturation increased the water permeability by about 100 times that of a specimen that had not been oven dried due to the resultant microcracking. A similar detrimental effect of gradual drying to 79% RH was reported by Powers et al. (1954). This highlights one of the important variables in water permeability measurements. Pressure permeability is relatively difficult to measure accurately and with repeatability. It also requires specialised equipment.

The most commonly used pressure penetration procedure is BS EN 12390-8 (based on the previous DIN 1048 procedure). Compliance tests on one grade of high performance concrete measured a range of water penetration depths from 0 to 12 mm, approximately 70% of the approximately 150 values reported by an independent laboratory were 0 mm. Pocock and Corrans (2007) for a different grade of concrete reported a mean water penetration of 8 mm with a range of 30 mm and a coefficient of variation of 125%. The target mean penetration depth was −6.5 mm. Concrete Society Technical Report 31 refers to a study in the United Arab Emirates where the coefficient of variation for 399 results was 65%.

These limitations suggest that the pressure permeability tests are not particularly suitable for durability performance specification and certainly not ongoing compliance testing.

7.6.5 Air permeability

Swiss Standard SIA 262:2003 "Concrete Construction" states: "The impermeability of the cover concrete shall be checked by means of permeability tests (e.g. air permeability measurements) on the structure or on core samples taken from the structure". The PermeaTORR, developed by Roberto Torrent, is an instrument designed precisely to measure air permeability of the cover concrete on site. The method serves to measure the coefficient of air-permeability of the cover concrete on site, in a non-destructive manner, and operates as described next (also see Figure 7.15). Vacuum is created inside the two-chamber vacuum cell, which is sealed onto the concrete surface by means of a pair of soft rings, creating two separate concentric chambers. At a time between 35 and 60 seconds (with a vacuum of about 5 to 50 mbar, depending on the concrete, instrument, etc.), Valve 2 is closed and the pneumatic system of the inner chamber is isolated from the pump. The air in the pores of the material flows through the cover concrete into the inner chamber, raising its pressure (P_i). The rate of pressure rise (ΔP_i; measurement starts at $t_o = 60$ s) is directly linked to the coefficient of air permeability of the cover concrete.

A pressure regulator maintains the pressure of the external chamber permanently balanced with that of the inner chamber ($P_e = P_i$). Thus, a controlled unidirectional flow into the inner chamber is ensured (Figure 7.15) and the coefficient of permeability to air kT (m²) can be calculated as described in the following equation. A microprocessor stores the information and automatically calculates the air permeability coefficient value kT (m²) that is displayed at the end of the test. The end of the test occurs when ΔP_i rises by 20 mbar or, in cases of highly permeable concrete, after 6 or 12 minutes, depending on the instrument's brand, from the initiation of the test.

Consequently, depending on the concrete permeability, the test may take from 2 to 6 minutes (12 minutes for one product). The microprocessor is capable of storing a great deal of test data and the information can be transferred to a PC for further analysis and filing. The function of valve 1 is to restore the system for a new test by ventilating it with air at atmospheric pressure.

Since the geometry of air flow is well defined, it is possible to calculate the coefficient of permeability with the following equation, derived in Torrent (2009).

$$kT = \left(\frac{V_c}{A}\right)^2 \frac{\mu}{2\varepsilon P_a} \left(\frac{\ln \dfrac{P_a + \Delta P_i}{P_a - \Delta P_i}}{\sqrt{t_f} - \sqrt{t_o}}\right)^2$$

where
  kT = Coefficient of air-permeability (m²)
  V_c = Volume of inner cell system (m³)
  A = Cross-sectional area of inner cell (m²)

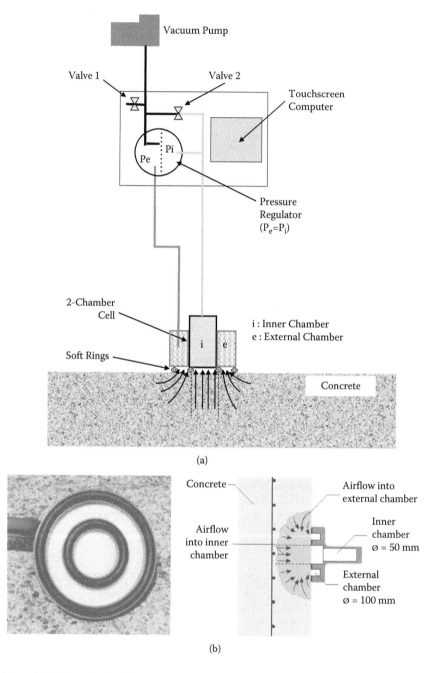

Figure 7.15 (a) A sketch of the test method. (b) Details of cell and air flow.

μ = Viscosity of air (= 2.0. 10^{-5} Ns/m^2)

ε = Estimated porosity of the cover concrete (assumed = 0.15)

P_a = Atmospheric pressure (N/m^2)

ΔP_i = Pressure rise in the inner cell at the end of the test (N/m^2)

t_f = Time (s) at the end of the test (2 to 6 or 12 minutes depending on the instrument brand)

t_o = Time (s) at the beginning of the test (= 60 s)

The knowledge of kT allows the estimation of concrete depth affected by the test (typically between 10 and 50 mm), which is also indicated by the device. The air permeability kT is very sensitive to the "covercrete" microstructure, covering some 6 orders of magnitude (0.001.10^{-16} to 100.10^{-16} m^2). Table 7.3 shows the classification of concrete permeability (ages from 28 to 180 days) as a function of kT.

The original Figg tests originated in the United Kingdom but have subsequently been neatly combined into a single instrument by James Instruments in the United States. A hole is drilled into the concrete (which may be in situ concrete or a test specimen) and a plastic plug inserted to create a cell below the surface of the concrete. A hypodermic needle is inserted through the plug to provide access. The first test involves applying a suction to the cell so as to draw in air through the surrounding concrete. The (very small) volume of air is measured by the movement of mercury in a tube through which the suction is applied. The second involves filling the cell with water and using movement in the same tube (but in the opposite direction) to measure the rate at which water is absorbed into the surrounding concrete.

The Wexham variant identifies two problems sometimes encountered with the aforementioned test. One is that air permeability is substantially affected by moisture content. The other is that air may be entering via defects in the concrete or a leaking plug rather than via permeable concrete. These two potential problems are solved, first by using a slightly larger diameter hole and including an instrument to measure humidity in the hole. Second, pressure rather than suction is employed so that any leaks can be detected by bubbles in a soapy water film on the surface.

An additional advantage of these kinds of in situ tests is that they can be used to measure the adequacy of curing (which has a large effect on

Table 7.3 Concrete permeability classes

Class	kT (10^{-16}m^2)	Permeability
PK1	<0.01	Very low
PK2	0.01–0.10	Low
PK3	0.10–1.0	Moderate
PK4	1.0–10	High
PK5	>10	Very high

penetrability). Potentially a contractor could be required to continue or resume water curing until an acceptable permeability is achieved.

7.6.6 Chloride diffusivity

Ions will naturally diffuse through the water-filled pores and microcracks present in concrete from areas of high concentration to areas of low concentration. This process will occur without any hydraulic gradient being necessary and thus is different from pressure permeability. Chloride diffusivity plays an important role in long-term performance, particularly after a high salt concentration has been established in the surface layer as a result of sorpivity.

Laboratory measurement of diffusivity used to be conducted using a diffusion cell. The time and equipment required to effectively measure diffusivity using this procedure limited its use to research applications.

Bulk diffusion tests, such as the Nordtest NT Build 443 or ASTM C1556, expose saturated concrete to highly concentrated chloride solution for a minimum of 35 days. Profile grinding and testing for chloride content enable the chloride diffusion coefficient to be calculated using Fick's second law. Therefore a chloride diffusion coefficient can be measured in two to three months. This may be suitable for verification of trial mix properties but not ongoing compliance testing. One appeal of chloride diffusion testing is that it can be used directly in service life prediction models based on Fick's law. An increasing number of specifications require chloride diffusion to be measured, often with the shorter coulomb test or migration test to facilitate compliance testing during construction.

Chloride diffusivity can reduce by orders of magnitude during the service life and therefore a single measurement of early age diffusion alone cannot predict the long-term performance. The improvement over time is primarily due to the composition of the cementitious binder. This is why many specifications require minimum replacement levels of fly ash or GGBS in an attempt to ensure certain improvement in diffusivity over time. However, prescriptive requirement on cementitious materials is contrary to the aim of performance specification. Thomas and Stanish (2003) measured chloride diffusion for periods from 90 to 180 days and 90 to 1550 days to establish the time-dependent effect as shown in Figure 7.16. Clearly projects cannot wait for 4 years to get a result but the research suggests a series of early-age tests could help confirm model assumptions.

Another limitation of bulk diffusion tests is the difference between actual chloride penetration over time and the measured chloride diffusion coefficient. Figure 7.17 shows the calculated chloride diffusion coefficient from the in situ chloride profiles after 19 years exposure and the chloride diffusion from NT Build 443 test on the uncontaminated concrete (Vallini and Aldred, 2003). The chloride diffusion from profiles is generally one to

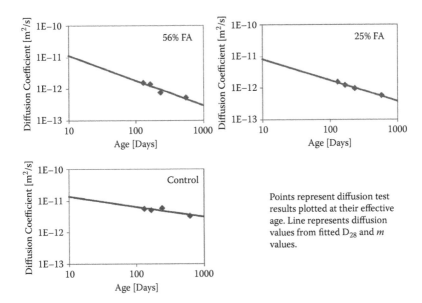

Points represent diffusion test results plotted at their effective age. Line represents diffusion values from fitted D_{28} and m values.

Figure 7.16 Comparison between fitted and measured diffusion coefficients. (After Stanish, K., and Thomas, M., *Cement Concrete Res.*, 33, 55–62, 2003.)

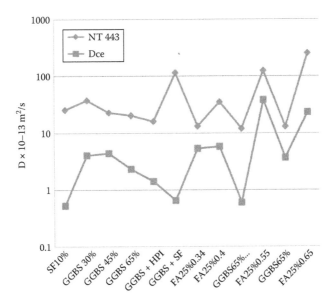

Figure 7.17 Comparison of diffusion coefficients from profile and NT 443. (After Vallini, D., and Aldred, J. M., Durability Assessment of Concrete Specimens in the Tidal and Splash Zones in the Fremantle Port, Coasts & Ports Australasian Conference, Auckland, New Zealand, 2003.)

two orders of magnitude less than from a bulk diffusion test. This appears to be due to the presence of other ions in seawater.

This field exposure trial included concretes with both low sorptivity incorporating hydrophobic admixtures and low diffusivity incorporating supplementary cementing materials. The results indicate that diffusivity is the dominant transport property under conditions of frequent wetting and limited drying.

7.6.7 Chloride migration

Because of the time to conduct a chloride diffusion test, there has been a great deal of attention to electrically accelerating chloride penetration. Nordtest NT Build 492 non-steady state chloride migration test adapts the ASTM C1202 resistivity test to rapidly measure chloride penetration. A 50 mm thick specimen is exposed to a potential difference depending on its resistivity. After a prescribed period, generally only one day, the specimen is split and sprayed with silver nitrate solution to determine the chloride penetration. Therefore it is quick as well as avoids the laborious profile grinding and chemical analysis necessary for the NT Build 443 or ASTM C1556 procedures.

Tang and Sorensen (2001) found that the chloride migration procedure has a repeatability coefficient of variation in the range of 5% to approximately 9%, and reproducibility in the range of 12 to approximately 24%. They conclude that; 'This test is therefore a good alternative method due to its simplicity, rapidity, good precision and fairly comparable results with the NT BUILD 443 test". As the migration test can be done very rapidly, it will generally give higher migration values than the diffusion test due to the greater average maturity of the test specimen in the latter. Tests conducted on concrete in Australia after 28 days curing showed the migration coefficient (NT Build 492) was roughly 70% greater than the diffusion coefficient (NT Build 443).

Some concrete technologists are passionately against electrical acceleration on the basis that it is totally artificial. However, ions are charged particles and therefore their electrical interactions and necessary charge balance is part of their movement. We would suggest that both diffusion and migration are measured before construction to determine performance and migration (or even resistivity) should be used for ongoing compliance.

Audenaert et al. (2010) measured chloride migration at 28, 56, and 90 days up to 5 years to determine time dependent effect. During verification testing, the authors would suggest that chloride migration testing between say 7 – 90 days might help eliminate the need for prescriptive cementitious material limits.

7.6.8 Resistivity

Resistivity can be easily and quickly measured on concrete. It is strongly influenced by moisture content and therefore the most popular resistivity test on concrete is conducted on saturated specimens. The ASTM C1202 test provides an electrical indication of the test concrete's ability to resist chloride ion penetration on vacuum-saturated 50 mm thick specimens. ASTM C1202 involves subjecting the concrete specimen to a 60 volt potential for 6 hours and measures the ease of which concrete allows charge to pass through it. This is affected by the density of the pore structure, the presence of insulating or conducting components and the availability of negative ions within the pore solution (Cao et al., 1996). Although concerns exist over the relevance of the test, there are abundant comparisons that show good correlation with chloride diffusion and other transport properties.

The test is often called the "rapid chloride permeability test", but it does not measure chloride movement or permeability. It is a saturated resistivity test. An often forgotten point in the test procedure debate is the important role that resistivity plays in the propagation phase of corrosion. In our opinion, if one is intending to use this test, a little bit more effort would enable both resistivity and migration to be measured using virtually the same equipment following the NT Build 492 test procedure.

As the ASTM C1202 or Coulomb test is a saturated resistivity test, the result can be obtained virtually instantly and does not require 6 hours. Simple bulk resistivity tests on compliance cubes or cylinders can be used to provide valuable resistivity data with very little effort or expense.

7.6.9 Corrosivity

Almost every test for durability assessment discussed in this section has been a penetrability test measuring one or more transport property of concrete. However, there are some methods of enhancing durability by electrochemical methods that do not necessarily change transport properties, such as corrosion inhibitors. ASTM G109 tests the effect of admixtures on macrocell corrosion in concrete subjected to wetting and drying with a 3% sodium chloride solution. As the chloride solution penetrates the 25 mm concrete cover by absorption and diffusion, the test measures both penetrability as well as any effect of increasing the chloride threshold level. The main problem is the time to complete the test that is equivalent to 150 coulombs of macrocell current in the control which can take over two years in a decent quality concrete. Lollipop tests with centrally placed reinforcement in a cylinder and monitoring corrosion with linear polarisation also measure both penetrability and corrosion inhibition. Increasing chloride concentration or reducing concrete quality may reduce the time required but may give misleading results. Another difficulty in confirming the benefit of corrosion inhibitors is the wide range of measured chloride threshold levels as discussed by Ann and Song (2007). Trejo and Pillai (2004) proposed an accelerated test for determining chloride threshold levels, which may be useful in quantifying the benefit of accelerators.

We feel that corrosion inhibitors can play a significant role in durability enhancement. Until widely accepted practical test procedures are available for confirming the increase in chloride threshold level, the full potential of this group of admixtures may not be achieved.

7.6.10 Assurance of long-term performance

Obviously the only true test for durability is the test of time exposed to a severe environment. Many products and technologies may have had promising performance in the laboratory but poor long-term field performance. Epoxy-coated rebar, which, based on laboratory data, was once thought to be the final solution to problems of reinforcement corrosion. Yet, Sagues et al. (2001) state that "damage from corrosion of ECR (epoxy-coated rebar) has continued to develop steadily in the substructure of five major Florida Keys bridges. Since the first indications of corrosion ~6 years after construction, damage increased at a rate of ~0.1 spall per bent per year until the present ~20 years age of the structures, with no indication of slowdown". Certain hydrophobic ingredients proposed for damp-proofing concrete were found to have reduced effectiveness or even to be leached from concrete over time (Aldred, 1989).

In the light of lessons from the past, it would be wise to insist that a certain performance requirement has been achieved over a period of not less than 15 years in an environment similar to that anticipated in the new structure or repair. This is a most demanding requirement in a rapidly changing world. On the other hand, the number of materials proposed as solutions to deterioration that have either not persisted over time or even given inferior performance, suggests such a bold step is necessary until a far better understanding of the processes of deterioration of both reinforced concrete and the proposed protective measure is achieved. It is clear that a material with a long history of proven success would justifiably inspire confidence.

Measuring resistivity/chloride migration from 7 days (or even 3 days) to say 90 days would help establish the expected time-dependent improvement in chloride resistance.

7.7 NONDESTRUCTIVE TESTING

With nondestructive testing (NDT) it is necessary to be particularly careful to clarify the objectives of the testing and the assessment of the results. Clearly the strength of the concrete in the structure is not necessarily the same thing as the potential strength (according to a standard compression test) of the concrete as it leaves the mixer or delivery truck. If it is not clear which of these is being sought, it is unlikely that the relative merits of different testing procedures will be correctly assessed.

From one viewpoint, the strength of the concrete in the structure is what really matters. However, even if this is accepted, we still have to consider whether what matters is the current strength of the concrete in the structure or its eventual strength. If the requirement is to assess readiness for early stripping or prestressing, or termination of curing protection, then the current strength is the more important. If it is the load carrying capacity of the structure or its durability, then the eventual strength will probably be more significant.

If the intention is to regulate the proportions of the concrete mix currently being produced, it is equally not obvious whether the potential standard specimen strength or the current actual strength in the structure is what matters. If considerations of eventual strength and durability in a particular structure require a 30 MPa (4350 psi) strength but construction efficiency requires 22 MPa (3190 psi) at 22 hours for prestressing, then the latter requirement will clearly rule. If day-to-day temperatures vary very widely (as they do in parts of Australia) then it could be necessary to supply concrete of 40 MPa (5800 psi) 28-day strength one day and 60 MPa (8700 psi) 28-day strength the next. Of course it is always possible that it is economically preferable to supply 60 MPa throughout, rather than complicate the situation, but this option can be ignored for the purposes of this example.

In the more usual case, a particular concrete mix will have already been assessed as suitable for its intended purposes and testing will be undertaken only to determine when any change takes place in that mix. In this case any extraneous factor that affects the test result, such as variable compaction of the test specimen, or variable temperature, either of the supplied concrete or of the specimen during curing, will add to apparent variability and so reduce the efficiency of the control process.

Assessing the above range of possibilities, it appears that the only case in which NDT testing could be considered as a total replacement for typical compression testing of standard specimens is where an early age requirement ensures such a large excess of 28-day strength that control of that strength is unnecessary. Even in this circumstance, standard testing may still be desirable if any problems are encountered, as otherwise it may be difficult to establish whether the problems are mix problems or usage problems. To some extent the decision would depend on the quantities of concrete involved since the cost of control measures may be to a large extent "per pour", whereas the cost of providing excess strength to avoid or reduce control is definitely per unit volume of concrete. Thus if a few small units totaling, say, 1 cubic metre of concrete per day were involved, it would be economical to use an excessively high strength and do little testing of any kind. However, if 200 cubic metres per day were used in floor slabs to be prestressed at an early age, both specimen testing and some form of in situ testing would be obviously justified.

An important consideration is that it is not only the accuracy of a test that matters but also its relevance and the accuracy of the assumptions made in evaluating it. For example, a test cylinder left on an in situ slab may give a very accurate strength but may have a very different maturity and therefore a very different strength to the slab itself. A pullout test on the same slab may be much more variable but at least it is measuring the actual strength. A standard test cylinder combined with a maturity (e.g., equivalent age) measurement of both the cylinder and the slab might be more accurate than the in-situ-cured cylinder and as relevant as the pullout test, but it does depend on the accuracy of the maturity–strength correlation and, for example, the compaction of the slab. An ultrasonic test would also be very relevant and may be quite repeatable and accurate but would be totally dependent on the strength–velocity relationship assumed, which would be affected by such factors as moisture content.

The reader is referred elsewhere to Bungey (1993) for further details of various NDT tests, but the authors certainly see a place for such tests in the overall control operation. Particular examples are pullout tests on suspended floor slabs prior to early stripping or stressing, and Schmidt hammer tests on freshly stripped columns. The latter is not a very accurate test (especially if used informally rather than according to the manufacturer's routine), but it is an extremely quick and cheap test that could be used on every column as it is stripped and would give early warning of any severe problems. It has even been suggested that the test could be worth performing even if the strength scale is not read. The implication is that the depth of indentation or even the sound of the impact would alert a daily user to any drastic problem. Day found this to be the case with spun concrete pipes, where sound was a good indication, and the process could be compared to tapping the wheels of railway carriages to detect cracks. However, a thorough examination by a U.S. university team (Telisak et al., 1991) concluded that in situ maturity determination was the most accurate criterion of early-age strength.

When regular NDT tests are carried out it is very desirable to enter the results in the control system for graphing and analysis alongside the other test data. Such action will soon establish the extent to which the variation of strength in the structure is a consequence of basic concrete variation.

A development pioneered by Dr A.M. Leshchinsky is that of using multiple techniques of NDT concurrently. The idea is that although the correlation of any one such set of test results with compressive strength may be upset by some influence (e.g., ultrasonic pulse velocity is greatly affected by moisture content), it is less likely that two or more different tests will be similarly affected. Therefore the use of two or more techniques will give more certainty of a correct assessment than any number of repetitions of the same type of test. This is a further illustration of a point previously raised, that is, the relevance of a test result may be even more important than its accuracy in many circumstances.

Another point of interest is Day's experience in the 1970s with the use of two standard ultrasonic testers: the UK Pundit and the Dutch CSI. Day conceived the idea of casting pairs of test cylinders instead of the conventional threes and using an ultrasonic test on these in place of a third early-age cylinder. The two instruments agreed on the ultrasonic reading, establishing that they were both accurately reading a fundamental property of the concrete; however, the readings did not correlate well with compressive strength. So ultrasonic pulse velocity (UPV) may possibly be as relevant as compressive strength in determining the quality of concrete, but it cannot be used to establish compliance with a compressive strength specification.

7.8 CONCLUSION

It can be seen that the question of which tests are worth doing, and how frequently and thoroughly it is worth doing them, is greatly influenced by the circumstances. The circumstances include the extent of the remaining variability and its sources, and also the assumptions made about the cooperativeness and trustworthiness of the concrete producer by the organisation imposing the control (which may or may not be part of the producing organisation).

Chapter 8

Mix design

Ken Day's specific surface/mix suitability factor (MSF) technique has been used in many countries over more than 30 years and has been the basis for "instant" mix designs given over the telephone for immediate use, with no more information than a sand grading, a verbal description of the appearance of the coarse aggregate, and the use to which the concrete was to be put (details later). The system still works and provides a necessary requirement for the degree of cohesion (= sandiness) needed for a mix to avoid segregation at any required workability. However, the consequences of exceeding this necessary minimum are now less severe, since additional admixture rather than additional cement is now the remedy.

It had been intended to eliminate this technique from the new edition, but several people, Roberto Torrent prominent among them, invited to an informal review of the chapter protested that they still used the technique to at least provide a lower limit to mix fineness and, in some cases, as a basis for correcting mixes when change occurred in the grading of one or more of the constituent materials of a mix in current use. The method is therefore still provided toward the end of this chapter.

Several other factors complicate the mix design problem to the extent that the latest edition of Francois de Larrard's excellent work *Concrete Mixture Proportioning* runs to 400 pages. It is our intention here to reduce the necessary theory to a relatively few pages by combining theory with a degree of experimentation. It is contended that the process of mix design needs to be integrated with that of quality control of both incoming materials and the resulting concrete, and that even small producers need to remain alert to both positive and negative changes in the situation (i.e., to future technical developments). Two consequences arise from this contention. One is that every producer needs to either establish his own trial mix facility or reach an agreement with an independent lab, so as to stay amenable to offers of alternative materials and, subject to the provision of encouraging data by the offering supplier, prepared to carry out trial mixes, preferably actually used in a noncritical location. The other consequence is that the day when it might have been useful for purchasers of concrete to prescribe mixes is definitely past. There may be some materials or combination of materials that the (relatively rare) technically advanced specifier knows to be unsatisfactory or particularly beneficial, but, this excepted, it is undesirable that the prospective purchaser should inhibit the process of optimisation or the selection of material suppliers by the concrete producer.

A major advantage of this proposal is that a producer would develop a standard range of mixes for all purchasers with the same requirements. This would be of substantial assistance in achieving good quality control and enable the producer to economically provide a more complete range of tests. It would also enable the producer to reach cooperative arrangements with suppliers of materials for them to carry out control testing on their materials and give the concrete producer advance warning of change. Specifiers and purchasers should also be aware that the field of such materials is in a period of rapid change and that national standards and the like may well not be up to date on all possibilities.

The first step in designing a range of mixes is to select the materials to be used, but these will depend on the available production facilities. There may be only provision for a single coarse aggregate, in which case it is likely to be a graded 20 mm material, or there may also be provision for a second aggregate of 14 mm or less maximum size. Since coarse aggregate will constitute the largest part of the concrete, price will be a consideration, but minimising the requirement for the mortar fraction will be even more important, since that material will be distinctly more expensive. The coarse aggregate will need to be sufficiently strong for the highest strength concrete required, and to have satisfactory bond characteristics and not exhibit moisture movement. It will also be important that the production facilities for the material are such as to ensure a consistent quality, including grading, particle shape, and bulk density. Rounded gravel will probably provide the highest bulk density (by volume), but its bond characteristic will be important and should be checked by the indirect tensile (splitting) test, especially where high strength is required.

If there are two coarse aggregates they will be proportioned to give the minimum percentage voids. It will be some benefit that their relative proportions can be adjusted if there are changes in the grading of either material. The smaller aggregate alone may be preferable for very high strength concrete (over 200 MPa). If there is a choice of fine aggregate, this will probably be made on the basis of a flow test (see Chapter 3), which is influenced by the grading and particle shape and surface texture. If a satisfactory natural sand is not economically available, a crushed material can be considered and, where an available natural sand has been discarded as too fine (or perhaps too expensive) a combination of the two may be worth consideration. Crushed sands tend to have what is sometimes described as a "hollow" grading, with a shortage of medium-sized particles, and this may combine well with a fine natural sand. These days there may be a substantial difference between the quality of available crushed sands with mountains of discarded material from coarse aggregate production typically available almost free and high-quality material produced by specialised Japanese equipment, which is probably worth the additional cost.

Historically, there has been a great deal of emphasis on combined aggregate gradings conforming to published grading curves of percentage passing the range of sieves. Although this is not now considered to have as much importance as it once did, it is worth inspecting a graph or table of individual percentage retained. If this reveals a severe deficiency on one or more sieves it is very likely to result in a tendency to segregation, especially where high workability is required. It is also undesirable that there should be a substantial concentration on one or two consecutive sieve sizes, as this would inhibit particle packing. Both natural and manufactured sands will need to be checked for organic material (clay). Such material is not as certain to be deleterious as has previously been assumed, but certainly needs to be checked for its effect on water demand and especially any deleterious effect on admixture performance (since some organic clays can absorb admixtures [see Figure 3.8]). It is also possible that clay will affect bond to coarse aggregate. The proportion of sand or fine aggregate can be determined by one of two methods. One is to simply carry out bulk density/percentage voids tests on a range of proportions (without cement or water) to find the minimum percentage voids in the combined aggregates. The other is to use Day's MSF technique to match the overall mix cohesion to the intended use of the concrete. In fact it is probably worthwhile to do both of these (since the MSF check is a quick and simple calculation). The MSF check should be regarded as ensuring that there is sufficient cohesion in the mix for the intended use, but an excess is not seen as unduly wasteful, since it only results in an increased admixture requirement, the cost of which may be partly recouped by reduced placing cost (which may be the subject of negotiation).

Having established the aggregate proportions, it remains to determine what Boudewijn Piscaer would describe as the "powder content". The minimum amount of this is clearly the void space in the combined aggregates, but its composition will depend on the required properties of the concrete. These properties will depend on the required minimum strength, perhaps at a particular age, and perhaps also on the required permeability/durability, heat generation, and shrinkage—and certainly on the required workability.

The days when the composition of the paste was obtained by looking up a table of 28-day strength against w/c ratio are hopefully gone. These days the water content is essentially a matter of choice within reason. A typical value is between 150 and 180 liters per cubic metre of the concrete, with a lower value being chosen for higher performance concrete. The desired workability will be obtained by increased admixture dosage rather than a higher water content.

The powder content will be composed of cement plus a selected percentage of one or more of fly ash, ground-granulated blast-furnace slag (GGBS), rice husk ash, silica fume, metakaolin, and finely ground limestone. Entrained air may also be required for frost resistance or to assist in workability. The choice between these materials will be strongly influenced by relative cost and availability (which will differ very markedly in different parts of the world), and by the perceived or specified need for durability and for strength at various ages. In many, but not all, parts of the world, fly ash (PFA) or limestone will be the most economical solution, but GGBS will give greater durability and silica fume higher strength (at all ages). Perhaps surprisingly, superfine limestone also gives good early strength. Reliability of supply of consistent material and of prompt advice from the supplier of any variation will also be an important consideration for the concrete producer.

The aforementioned process may be regarded as simple or complex depending on the range of materials economically available and the perceived competence of the material suppliers. It is likely that a continued program of trial mixes will be worthwhile (largely on a production scale with the concrete supplied for a less critical use). The resulting mix may even be suitable for self-compacting concrete (the use of which is likely to increase substantially in the future) with an increased dose of high-range water-reducing admixture (HRWRA). This can be checked by filling an upside-down slump cone with it and lifting the slump cone. To be self-compacting the concrete must flow out to a radius of at least 550 mm. Although flowing outward it is important to observe the leading edge of the flow, this must be blunt, containing coarse aggregate. Even a slight "halo" of preceding fluid paste without aggregate indicates that the mix has insufficient fines to provide the necessary cohesion.

8.1 MIX SUITABILITY FACTOR (MSF)/SPECIFIC SURFACE THEORY (FROM THE THIRD EDITION)

The basic concept of specific surface (SS) mix design is extremely simple but requires modification to work effectively. The simple basis is that a given degree of workability will require an appropriate specific surface to avoid segregation, the higher the workability, the higher the required specific surface. Knowing the individual specific surfaces of the coarse aggregate and the fine aggregate, the required sand percentage can be calculated (see Table 8.1).

It is well known that a finer sand will have a higher water requirement than the same amount of a coarser sand, but specific surface theory says

Table 8.1 MSF values

MSF	Slump range		Remarks
	mm	Inches	
<16			Unusable, too harsh
16–20			Harsh mixes, only suitable for zero slump concrete under heavy vibration
20–22	0–50	0–2	Hard wearing floor slabs, precast products under good external vibration
22–25	50–90	2–3.5	Good structural concrete
25–27	80–100	3–4	Good pumpable concrete, fine surface finish, heavily reinforced sections
26–28	90–120	4–5	Pumpable lightweight concrete
27–31	>200	>8	Flowing superplasticised concrete
≥33			Self-compacting concrete—slumpflow ≥ 550 mm, T500 ≥ 2 seconds

Note: Add "33+ minimum flow diameter 600 mm", "T500 time 2–7sec" self-compacting.

Table 8.2 Modified specific surface values

Sieve fraction	Author's modified SS values	Approximately true specific surface $(cm^2/gm)^a$	Surface modulus
>20 mm	2	1	1
20–10	4	2	2
10–4.75	8	4	4
4.75–2.36	16	8	8
2.36–1.18	27	16	16
1.18–0.600	39	35	32
0.600–0.300	58	65	64
0.300–0.150	81	128	128
<0.150	105	260	256

[a] According to B.G. Singh (1958).

that, within wide limits, if the proportion of fine sand is reduced so that the specific surface of the combined aggregates is the same as with the coarser sand, the same water requirement and the same degree of cohesion will result. The original SS theory did not work in practice because it was found to overestimate the effect of very fine particles. The surface area of a sphere approximately doubles as its diameter halves, giving rise to the second column of figures in Table 8.2 (neglecting particle shape). Day's modification recognises that as diameter reduces, a point is reached where it takes less water to fill the voids in the material than it does to coat its surface. On a purely empirical basis, the first column in Table 8.2 "Modified Specific Surface", was originated by Day in the 1950s to implement this concept.

It was assumed at the time they were originated that these values would require subsequent refinement, but in spite of attempts to improve them in the laboratory, and by their use for production concrete in many countries, the figures have remained substantially unchanged for 50 years. However, it is now recognised that superfine aggregate particles can actually displace water from between cement particles. This can only happen in the presence of a HRWRA, as otherwise the fine particles will clump together and then will increase water requirement.

It would be more correct to use surface area per unit solid volume than per unit weight, but the weight basis was been retained because the actual numbers were familiar to users of the original SS theory. For the same reason, Day's original modified figures have been doubled so that the overall combined aggregate SS is of the same order as the original. However, where there is a large difference between the specific gravity (sg) of coarse and fine aggregates an adjustment is desirable.

Modification of the basic SS values is not the only adjustment required to make SS mix design work. Other factors to be taken into account include the following:

1. The effect of cementitious materials and entrained air
2. The effect of particle shape
3. A requirement for continuity of grading
4. Limitation of fineness and coarseness of sand grading

Before discussing these points, some of the objectives of mix design should be reviewed. Generally a sandier mix will have a higher degree of cohesion and be easier to handle and place. However, it will have a higher water requirement. Traditionally, water/cement (w/c) ratio has been regarded as the best criterion of quality, so that a sandier mix will require more cement and so be more expensive. Further investigation has shown that additional water is more deleterious than less cement at a given w/c ratio, increasing the desirability of minimising water requirement. So the objective of mix design is to achieve acceptable fresh concrete properties at minimum water content. With the advent of self-compacting concrete, the task becomes even more critical.

8.1.1 Effect of cementitious materials and entrained air

These materials increase cohesion and so reduce the required SS of the aggregates. Ken coined the term MSF (Mix Suitability Factor) to represent the combined effect of all constituents on cohesion. The formula is

$$MSF = SS + 0.025EC + 0.25(air \% - 1) - 7.5$$

where
> SS = Modified specific surface of combined coarse and fine aggregates
> EC = Equivalent cement content (see later)

8.1.2 Effect of particle shape

An intrinsic assumption in SS mix proportioning is that a finer sand will cause less disruption to the packing of the coarse aggregate, permitting a reduction in sand percentage. It is not necessarily obvious that this reduction is exactly the same as the reduction needed to maintain the same combined specific surface of the combined aggregates but this seems to work in practice. A more angular particle shape of the coarse aggregate also causes an increased requirement for sand, since it increases the percentage voids in the coarse aggregate to be filled by mortar. An increase of up to 3 in the appropriate MSF may be needed depending on the degree of angularity (which has a larger effect than flakiness or elongation). The actual surface area of both coarse and fine aggregates is obviously increased by a more angular particle shape at a given grading. However, whereas an increased fineness of a sand can be fully compensated by reducing its percentage (so there is no increase in water requirement), this is not so for a more angular fine aggregate since it does not reduce the interference with coarse aggregate packing, and may even increase it. So the angularity of the fine aggregate is neglected in determining the percentage to be used, but the predicted water requirement may increase by 5% to 15%.

Specific surface cannot be the only criterion for mix proportioning because it does not take into account particle shape and provides no assurance of continuity in the grading, which may be needed to avoid segregation and achieve pumpability. This is the aspect better covered by the void-filling theories, but Day believes he achieves a simpler and more workable solution by using crude, semiempirical corrections for these purposes.

8.1.3 Grading continuity

In the past, a great deal of research effort has gone into the search for an ideal aggregate grading. This has been to some extent pointless because, even if it exists, such a grading may be impossible or too expensive to attain with the materials available. One still sees requirements for sand grading to be within certain limits (particularly in the United States), but the move to abolish them is gaining momentum.

However, it is undeniable that gaps in an aggregate grading, while they may make the concrete easier to compact under vibration there is an increase in the tendency of the concrete to segregate. Resistance to segregation is vital in higher slump and pumped concrete. Gaps (differences) in excess of 4% to 5% between the percentage retained on consecutive sieves finer than 7 mm should be avoided if possible, especially if any tendency to segregation has been noticed.

8.2 LIMITATION OF FINENESS AND COARSENESS OF SAND GRADING

A wide range of sand fineness can be accommodated by appropriate adjustment of sand percentage to give a desired combined aggregate specific surface, but there are limits.

8.2.1 Upper limit of coarseness

A sand reaches the upper limit of coarseness when there is insufficient paste (cement, water, and entrained air) in the mortar to provide adequate lubrication. This occurs not so much due to the coarser sand requiring more paste per unit quantity of sand, but rather because more sand must be used to provide the desired surface area if it is coarser. If the sand quantity is not increased, the overall mix will be too harsh, and will segregate unless of very low slump. If it is increased beyond the limit, the water requirement rises to provide the required total paste volume required. Strength will be reduced, the concrete will almost certainly bleed severely, and workability will suffer in a different way, that is, it will have unsatisfactory mortar quality rather than an inadequate amount of mortar. A comprehensive mathematical treatment of this problem is given by Dewar (1999), but here we will deal only with a few rules of thumb. What is important is that users should recognise the problem when they encounter it. As noted earlier, this will not occur at a particular sand percentage for all mixes but will depend on several other factors. Some rules of thumb to indicate when the problem should be considered are

1. Sand percentages in the range of 50% of total aggregates (in low cement mixes) to 65% (in high cement mixes) (very rough guide).
2. Solid volume of sand exceeding about 5 times the solid volume of cementitious material. With normal sand and cement this can be taken as a sand to cement ratio of about 4 by weight. When fly ash or very heavy or light sands are involved, the volume figure applies. This guide is still not invariably accurate because the limit is affected by the particle shape and grading of both the sand and coarse aggregate and by the use of air entrainment.
3. From a different viewpoint, the problem may arise when the FM (fineness modulus) of the sand exceeds 3.0 in low cement content mixes or 3.5 in high cement content mixes. In ConAd specific surface terms the danger signals may be around 40 for high cement contents and 45 for low cement contents.

8.2.2 Upper limit of fineness

The fine limit for a sand is reached when a further reduction in sand proportion will leave insufficient mortar (sand plus cement paste) to provide adequate lubrication to the coarse aggregate. With a very fine sand it is possible to get quite close to using a cubic foot of coarse aggregate by loose volume in a cubic foot of concrete, and the shape and grading of the coarse aggregate makes a substantial difference to where the limit is. The limit will certainly be close; however, when the coarse aggregate approaches 60% by solid volume of the total concrete. Again from the other point of view, the problem is likely to arise with sands of FM around 1.5 (with a high cement content) to 1.8 (with a low cement content) or, in ConAd SS terms, in excess of 90 with any cement content. It is also possible that a high cement to sand ratio is intrinsically undesirable in the same way that a heavily oversanded mix is undesirable (e.g., higher shrinkage). A sand weight less than the weight of cementitious materials should be viewed with suspicion and avoided if possible.

8.2.3 Coping with extreme sand gradings

The important point is rarely the establishment of the exact limit, rather it is the fact that within these quite wide limits, grading is not the problem that most typical specifications would suggest. It is of course necessary to accurately determine what proportion of sand should be used in each particular case and this is the main strength of the method of mix design evolved by Day.

An example of the coarse limit was encountered in Indonesia. The local sand on occasions had less than 3% passing a 300 micron sieve. Its fineness modulus was only of the order of 3.0, which did not seem an excessively high figure. However, its specific surface of 40 to 42 was clearly excessively low. Increasing the proportion of this sand did not solve the problem, which was excessive bleeding. Eventually a choice had to be made between a proportion of finer sand, even though not locally available and so very expensive, and the use of additional cement purely for bleeding suppression. Another alternative would have been air entrainment, but this was rejected, again due to nonavailability locally and also because the production personnel were unfamiliar with it and had no test experience or equipment. There have been very coarse sands in Singapore and in Australia requiring 48% to 55% of sand, but these have all occurred when relatively high cement contents were required. In an extreme case, where the sand is very coarse and only a low strength and therefore a low cement content is required, the following possibilities should be considered:

1. Use of a small proportion of a second fine sand (even if quite expensive).
2. Use of a small proportion of crusher fines with a high "fines" content.

3. Use of fly ash, which has 37% greater volume than an equal weight of cement (if in an area where fly ash is inexpensive, more might be used than strictly necessary for strength).
4. Use a proportion of GGBS or of superfine calcium carbonate.
5. Use of air entrainment (as valuable, volume for volume, as cement for this purpose).
6. If no alternative is less expensive, the use of more cement than necessary on strength grounds would certainly solve the problem since it both reduces the sand percentage required for a given MSF and provides more paste to fill the sand voids. However "cement" these days, as noted earlier, is likely to be a composite material.

Extreme testing of the fine limit has also occurred. In 1956 (Day, 1959) a case was encountered where the sand percentage calculated by Day's system came to 15% (virtually all the sand passed the 300 micron [No. 50 ASTM] sieve). It proved possible to obtain a ¼ inch (7 mm) single-sized crushed rock and the concrete was made with 10% of this material and 15% of sand (the balance being 75 percent of an almost single sized 20 mm [3/4 inch] crushed rock).

During the early development of the system (in the early 1950s in England) sand percentages of 22% to 23% were used, but although the sand was purchased as "plastering sand" rather than "concreting sand", this was an example of the use of a very low "MSF" on earth dry concrete rather than the use of a very fine sand. It should always be possible to use a proportion of crushed fines (choosing a coarse variety) when the natural sand is too fine for use alone. However, the particle shape of the crushed fines will increase water requirement, and therefore increase cement requirement, at least somewhat.

In selecting all constituent materials for concrete it is particularly important to take into account consistency of supply. Any variation in the characteristics of a material is likely to cause variation in the resulting concrete, unless the concrete producer has been given advance warning and been able to make a compensating mix change. It is bad enough that the concrete producer has to maintain skilled staff able to make rapid mix adjustment, worse still if he has to be able to detect change without advice from his supplier, but worst of all if change is undetected or inadequately compensated for increasing variability and so requiring a higher mean quality, presumably at higher cost. Hopefully the adjustment will be timely enough to avoid failures or penalties.

Day's SS of an aggregate differed from true specific surface because he recognised that as the particle size of the finer sand/aggregate fraction reduced a stage would be reached where less water was required to fill the voids in the mass of aggregate than to provide a surface coating of water. It is now clear that it is not sufficient to recognise the assumed reduction in water requirement for this reason, because the finer particles can actually displace water from between cement particles, reducing water requirement in the same way as does fine fly ash, calcium carbonate, and so on.

Apart from the increased range of supplementary cementitious materials, the other major change has been an increased range of more powerful water-reducing admixtures, to the extent that any degree of fluidity can be produced with almost any selected quantity of water. Therefore an increased MSF (i.e., fines content) no longer has to be paid for in increased water and cementitious content, although the admixture cost will increase. Especially where sand/fine aggregate is less expensive than coarse aggregate, sand contents unthinkably high on the old basis can now be used with no concern that fresh or hardened properties, including strength, permeability, and shrinkage, may be affected.

A new aspect of mix design is attention to the elimination of internal voids. The initial voids figure can be determined by subtracting the volume of the coarse and fine aggregates from one cubic metre. This volume then desirably has to be filled, perhaps with a small excess, by the combined volume of water, cement, entrained air, admixture, and fine material such as fly ash, GGBFS, silica fume, rice husk ash, or superfine calcium carbonate. If there is an inadequate volume of such superfine material, it may be necessary to include a VMA (viscosity modifying admixture) to avoid bleeding and segregation, especially where high workability or good pumpability are required.

Since the last edition Day has designed only one mix and it is worth briefly recounting. Using the University of Texas ICAR program recommendations (incorporating the aforementioned concepts), Day took the aggregates of an existing normal workability mix and in a site laboratory increased sand content in small increments until a minimum voids content was found at a little under 50% sand (bulk density less the weights of coarse and fine aggregates each divided by their SG gives %voids). He then reduced the water content from 170 to 150 liters/m3 and added superfine limestone until the volume of cement plus fly ash plus limestone plus water equaled the volume of voids in the aggregates. Finally a high-range water reducer was added to provide very satisfactory, fully self-compacting concrete (SCC), concrete of slightly higher strength and lower shrinkage than the original mix. OK, you can do that, but you will not necessarily have the most economical solution.

There is now a large variety of cement replacement and other fine materials, and there may be variation in characteristics and quality of the same material from different suppliers. Other than in broad principle, as set out by Barry Hudson, or in the detail involved in Francois de Larrard's treatise (1999), such variations are not amenable to theoretical or numerical mix design and their relative merits must be established by trial mixes.

To be at their most competitive, concrete producers need to either establish their own trial mix facility or reach an agreement with an independent lab to assess offers of alternative materials. Subject to the provision of encouraging data by the offering supplier, the concrete producer can carry out trial mixes, preferably actually used in a noncritical location. There may be some materials or combinations of materials that a technically advanced specifier knows to be unsatisfactory in some respect but, this excepted, it is undesirable that the prospective purchaser should inhibit the process of optimisation or the selection of material suppliers by the concrete producer, although he may reasonably require a wider range of test and performance records of a proposed mix before approval.

A major advantage of this proposal is that a producer would develop a standard range of mixes for all purchasers with the same requirements. This would be of substantial assistance in achieving close quality control and enable the producer to economically provide a more complete range of tests. Specifiers and purchasers should also be aware that the field of such materials is undergoing a period of rapid change and that national standards and the like may well not be up to date on all possibilities.

But now it is time to get Hudson's more detailed advice. For this edition Hudson has contributed his methods of mix origination. Hudson is currently responsible for operational performance of 16 million cubic metres of concrete per annum produced in 485 plants in 12 countries, so he is worth taking notice of.

Obviously, there are many different methods of concrete mix design in practice today. These designs range from the very basic 1 shovel of cement, 2 shovels of sand, and 4 shovels of coarse aggregates, through to some sophisticated software programs that require proprietary test methods and results to determine constituent proportions.

In this chapter, we will discuss a successful method of concrete mix design that does not rely on individual constituent characteristics, but rather how a group of materials perform when proportioned together for a given production facility or batch plant. There are many methodologies for designing concrete mixes. In reality, the scope or range of material proportions that can make a workable concrete are relatively narrow, with some broad rules of thumb and common sense setting some boundaries. In this chapter, we will not discuss high-performance or value-added concretes, but concentrate on the main volume of concrete that most commercial ready mix concrete plants will produce and deliver, the "vanilla" concrete (normal grade strength and workability concretes) (20–40 MPa, 50–150 mm slump). These concretes will typically consist of one or two sands, anywhere from one to six coarse aggregates, a cement (maybe ordinary Portland cement or a cement that has some supplementary cementitious materials (SCMs) such as fly ash or slag added). Some admixtures (normally water reducing and/or air entraining) and of course water.

The objective when making these vanilla or normal concretes (15–30 MPa, 50–150 mm slump) is typically to meet some type of compressive strength requirement; to have an agreed workability; be accepted by the pump/placing contractor; and not to bleed, segregate, shrink, and so forth.

Other constraints can be placed on the material proportions that are eventually used in production. These typically the cost of materials and, increasingly, the reduction in energy or the CO_2 emissions required to make the concrete.

As mentioned at the beginning of this chapter, in reality the range of the various material proportions to be used in concrete is relatively narrow. And there are very few materials to proportion.

8.3 PROPORTIONING AGGREGATES

The aggregate fraction of the concrete mix will have two distinct fractions: (1) fine aggregates or sand (usually aggregates smaller than 5 mm), and (2) coarse aggregates (greater than 5 mm). Rather than performance or practical considerations, the main criteria in selecting these sands and aggregates are usually to conform to a prescribed grading or particle size distribution specification. This requirement is usually regardless of whether the aggregate is natural sand and gravel, or a quarried material such as limestone, granite, or basalt (see Table 8.3). Coarse and fine aggregates will make up by far the majority of the volume of a concrete (between 70% and 90% of the total volume). There is a lot of debate on how to proportion coarse and fine aggregates in a concrete mix, and also within themselves. For example, should the coarse aggregates be proportioned to get a continuous grading curve? Or, should they be proportioned to have minimum voids (voids being the measure of air in a given bulk volume of aggregates)? Should the sand be proportioned to achieve a target FM (see Chapter 3, Figure 3.2) or

Table 8.3 Influence of aggregate shape characteristics on hardened performance of concrete

Aggregate	Compressive strength (MPa)	Flexural strength (MPa)	Flakiness	Elongation	Angularity number
Natural quartzite gravel	40.7	3.65	8	24	1
Crushed limestone	44.8	5.48	26	33	7
Crushed basalt	47.6	5.10	31	42	9
Crushed flint gravel	37.8	4.55	34	42	9

to have the fastest flow time in a sand voids test? Should these materials be proportioned for some other characteristic, like surface texture, absorption, density, or price?

In reality, there are many properties of aggregates that influence either the plastic or hardened properties of concrete. In truth, it is difficult to point to any change in concrete performance, whether in the concrete plastic or hardened state, where that change can be directly correlated to a change in an aggregate characteristic (e.g., grading, shape, texture, surface texture, LA abrasion, Micro-Deval, sulfate soundness) other than durability issues due to contamination or reactive components. There will no doubt be a change in concrete performance when one or a combination of these characteristics change, but it will not be possible to predict exactly what the magnitude of the change will be, especially over a wide range of aggregates and concrete mix designs.

Having decided on the blend of the aggregates, the basic deviations that may change the ideal blend must be considered. For example, variation in aggregate quality, which can be extreme. Of course, it is necessary to take into account the accuracy and capability of the concrete batch operation. Therefore, in proportioning these coarse and fine aggregates and the other materials in the concrete, the material consistency and the capability of the production process combined must ensure the robustness to be able to deliver a consistent product to the placer and the specifier. The changes in aggregate or cementitious quality, combined with production variances, need to be absorbed by a mix design so that each change does not produce a noticeable change in the concrete properties for the placer or finisher, and wide variances are not experienced in the concrete quality control data as covered in Chapter 10.

8.4 MATERIALS PROPORTIONING

As mentioned earlier there is a narrow workable range of concrete constituent materials when the materials are proportioned. In general, the workable percentages for the amount of sand in the total aggregate matrix (the combination of all sands and coarse aggregates) are between 30% and 60%. These percentages assume that you are working with usable quality concrete sands and aggregates.

Typically anything less than 30% of sand in the aggregates blend will yield a concrete that is very harsh or boney; it will probably have segregation issues and will not be cohesive. Running the concrete through a pump would be difficult. Likewise concrete where the sand percentage is over 60% typically requires a lot of water to attain a given workability; the concrete will be prone to bleeding and often a high degree of shrinkage occurs as a result. From these rules of thumb, the workable range of coarse aggregates in a concrete mix will be between 40% and 70% of the total aggregate volume.

Once the aggregate blend has been identified, the remaining two key ingredients to complete the matrix are water and binder. (In this example we will assume we are using a single Ordinary Portland Cement).

One of the oldest rules of concrete is that the compressive strength of the final concrete will have a relationship to the water to cement ratio that is used to make that concrete. It may come as a surprise to many, but when the solid constituents that make up the vanilla concrete mix design (sand, coarse aggregate, and cement) are mixed together in the workable proportions noted earlier, the water required to take the matrix to its plastic limit (the point at where the matrix moves from a semisolid to a plastic state) doesn't change a great deal from very low cement contents ($180 \ kg/m^3$) to very high cement contents ($700 \ kg/m^3$). The total change in water required to reach the plastic limits of the matrix of materials will only change by approximately 10% across the workable range of concrete designs, with a higher cement content resulting in the slightly higher water demand.

As a quick check, the absolute volume of water per cubic metre or yard that is used in a batch plant should be within this 10% range. Obviously, there will be some outlying cases where this is not true, but for most normal concrete this should be the case. We now know that of the four basic constituents (sand, coarse aggregate, cement, and water), one constituent (water) is going to be fairly constant. We also know that the compressive strength should be a function of the w/c ratio. If there is a prescribed cement content or w/c ratio, and the absolute volume of water required is known, two of the four components of the mix design have their volumes determined.

For example, if there is a specification that requires a minimum 350 kg cement per m^3, and we know that the volume of water required to make the given constituent materials plastic is 170 kg, then 520 kg of the mix design is determined. Alternatively, if the required mix design needs to have a maximum w/c of 0.50, the amount of cement required to make the matrix plastic will be $X/170 = 0.50$, therefore $X = 340$ kg.

So the question is how to ascertain the amount of water required to get the matrix of materials plastic. There are two very simple methods. The first is where the materials are available in a laboratory situation. We would suggest combining for simplicity the coarse and fine aggregates together 50/50 either by weight or by volume. Add to the mix of aggregates a midrange quantity of cement determined by the cement content that is currently use for a given compressive strength, or if dealing with new materials pick a cement content of say 300 kg/m3. Add these materials to the mixer and then add water until the concrete just becomes plastic. (The state at which the concrete has become plastic if you were measuring by a slump cone would be just "wet" enough to have some slump, no more than 10 mm or ½ inch.) Calculate the amount of free water (based on saturated surface dry aggregates) added to the matrix to get the concrete to its plastic state. This will be the volume of water that you will base your mix designs on for this set of materials.

When designing concrete, the saturated but surface dry value for an aggregate should be the mass that is used as the constituent mass for a given volume of concrete. The reason for this is relatively simple. It is assumed that when making a batch of concrete that any water added to the batch will be absorbed by the sands and aggregates (assuming that they are dry), or alternatively if the aggregates are wetter than saturated but surface dry, the aggregates will add water to the constituent batch materials.

Ordinary Portland cement (OPC) as a material has a very small range of densities and is an ingredient in the concrete mix design that is treated as dry. Typically, the density of OPC is in the range of 3.12 to 3.15 t/m^3.

Water, unless otherwise specified, is treated as having a density of 1.000 t/m^3, and air is treated as having a density of zero. However, when calculating the volume of concrete, it is necessary to nominate the volume of air that is entrained or "entrapped" so that the constituents of the mix design can be proportioned so that when batched, mixed, and placed, the constituents result in a predetermined volume of concrete (typically in its fresh state.)

All constituents need to be accounted for in a concrete mix design as far as volume is concerned, which includes all additives and admixtures.

Understanding the volumetrics of all of the materials, the influence and calculation of the impact of free or absorbed water, and how to account for air, both entrapped and entrained, are fundamental to all mix design processes. However, in the experience of the authors, that is typically where the similarities in mix design approaches end.

Another method to estimate the volume of water required to get a set of materials into a plastic state is to study the production records or batch records from the concrete batch plant. Simply plot the free water added to the concrete (using the batch moisture contents of the constituent aggregates and added water, and adjusting for the estimated water reduction of any admixture used). This resultant water volume will be a very good guide as to the "water demand" of your set of concrete making materials. From the batch and quality control records, plot the relationship between compressive strength and water/cement ratio. From this plot, you can derive the cement content that is likely to be required to produce concrete of a desired strength, knowing of course that the water volume is relatively constant across the range.

The resulting volume of sand and aggregate, or coarse and fine aggregate needs only to have the blend or ratio of coarse to fine (in volume) determined. There are many determining factors that go into making a decision on what the blend or ratio of these materials will be.

Usually in plain concrete, the cost of the final blend of materials is an important consideration because typically in developed markets, the profitability or margins in producing and selling concrete are extremely low. This being the case, and aggregates typically comprising 15% to 30% of the total cost of producing a volume of concrete, one major consideration as to the blend will be related to the actual cost of the materials.

However, the cost of the aggregate is not the deciding factor on the total cost of the all of the materials that go into making that volume of concrete. For example, in general terms, you could afford to pay 5% more for a ton of sand if the more expensive sand was to allow you to use 1% less cement.

8.5 VOLUMETRICS

A lot has been written about volumetric mix design, particle packing, and minimum voids. When you have a collection of granular materials such as cements, sands, and aggregates, there is only one combination of all of these materials that will give you a maximum packing density. There are many aggregate blending programs and methodologies that will either allow you to calculate the combinations of materials to give you the maximum or indeed a targeted density; or, in reverse, a percentage of voids in the mix. The .45 power curve that is commonly used in the United States is one good example

It is the authors experience, however, that the maximum density, or minimum voids content of all of the materials, or even just the sands and aggregates will probably not give you a concrete that is workable or easy to finish, and it probably won't pump very well, if at all! And therefore it would not be commercially viable. The maximum density concrete made by having the lowest voids content or highest aggregate packing density will in many cases give a high compressive strength—in the laboratory.

Is it really an advantage to have the maximum packing density in our aggregates? We think not. The "best" concrete has aggregate packing densities that are lower than the maximum. It is the packing density of all of the materials in a concrete mix design that is important. A given packing density, or voids content can be achieved by many different combinations of the materials.

If you look at the ISO packing pyramid used by de Larrard and others, you will notice that the isobars allow quite a wide range in the combinations of the materials to be blended to yield the same result.

There are many characteristics that influence a material and how it packs, or that particular material's void content. These characteristics have differing impacts depending on the size and shape of the material in question. Factors that influence aggregate and sand packing efficiency are

Table 8.4 Typical contribution of aggregate properties to the performance of concrete (Kaplan)

Property of concrete	Shape	Property of aggregate texture	Modulus of elasticity
Flexural	31	26	43
Compressive	22	44	34

1. Size
2. Grading
3. Shape
4. Surface texture
5. Coatings

All these characteristics play a significant role in how a solid will behave in concrete, both in its fresh and hardened states. Table 8.4 shows the relative effect of different aggregate properties on flexural and compressive strength of concrete. Each of the characteristics influences a particles specific surface, but expressing the aggregates specific surface by a simple voids number is meaningless as that single number (voids) can be arrived at by many combinations of the individual characteristics.

Because coarse aggregates are being used as inert fillers in concrete, we suggest that getting a coarse aggregate that has the lowest void content (maximum packing density) and that has the lowest cost. A good range of packing density (voids content) for a coarse aggregate is 34% to 39% (tested to ASTM C1252 the test method). Of course, it is also important to take into account other properties such as maximum particle size, durability criteria, and any other constraint that may be specified for that aggregate. The aggregate also has to be a competent material that is fit for purpose. If an aggregate is already in commercial use as a concrete aggregate, you can operate under the assumption that the aggregate satisfies these requirements. If in doubt, however, have a reputable laboratory check the aggregate for suitability.

All blending of aggregates should be done on a volumetric basis unless the aggregates being blended have the same specific gravity.

To improve the packing density of an individual aggregate will typically require reprocessing the aggregate by some method, typically screening or crushing the aggregate through a crusher that can improve the particle shape. Especially in coarse aggregates, the two major influencers of packing density are both aggregate-process related. The two properties are particle shape and particle size distribution. The more equidimensional (closer to a sphere) a particle is, the higher its potential packing density will be (the lower the voids content). With particle size distribution (or grading) having an "even" grading will not result in the highest particle packing. Actually, a grading or particle size distribution where there are gaps in the particle distribution will give a higher packing density than that of evenly graded.

To really understand the aggregates, sand, and powders that you are using in concrete, the grading should be expressed as individual percentages retained on a given sieve or size measurement. This is particularly important when it comes to the finer particles. This is important because if you have too many particles that are in the same size range, their packing efficiency will be very low (similar sized particles pack together inefficiently leaving large void spaces that need to be filled by even finer particles) and this will have a detrimental impact on the quality of the concrete, or the concrete mix will require a higher water to cement ratio or more cement to reach a desired workability. This will obviously have a negative impact on cost, shrinkage, heat generation, and so on.

When considering volumetrics in concrete, the intention is to fit the finer particles between the voids left by the larger particles. Typically, the larger the particle in a concrete mix, the lower the cost per ton of that material.

The amount of paste required to make a plastic concrete mixture (in this case paste being cementitious materials, sand particles less than 100 micron, and water) equals the total volume of voids space in the aggregate matrix plus 3% or 4% more absolute volume to allow the system to "float", or become hydraulic.

In general, when you have the option to use more than one coarse aggregate in a concrete design, it will yield better results than using a single-sized aggregate. It is very difficult to try to get a single coarse aggregate to have an even grading (a grading that will give a low voids content) and not be prone to segregation. In the United States, the most common sized concrete coarse aggregate is a "57", which in metric units is a 20 mm–5 mm material.

The 57 grading covers too many sieve sizes and is prone to segregation, as well as being very sensitive to the inconsistencies in aggregate production. The result is a very inconsistent grading that can go from very fine to very coarse. Obviously, as the grading changes, so do many of the other properties of the aggregate. One such major change is the voids content of the coarse aggregate. Because the voids content on the coarse aggregate fluctuates, the volume of finer materials that are required to fill those voids (sand, cementitious, and water) will also change.

In the majority of cases where the mix design methodology being described here has been employed, and acceptable quality natural concrete sands and coarse aggregates (both natural and crushed) are used, the rough percentage of sand to that of coarse aggregate for everyday concrete (10–35 MPa, 50–200 mm slump) yields an ideal sand percentage (by volume) of around 55%, with that percentage decreasing to around 45% when you get into the high powder factor concretes that are typically deemed high-performance concrete (HPC) in excess of 500 kg/m^3.

HPCs will often contain fly ash or slag, and silica fume. To accommodate this extra volume of powder, the optimum volume of sand will decrease.

This may sound completely the opposite or counterintuitive compared with current practice in a mix design. But this point should be considered in the context of the method of mix design. Water is to be kept to a minimum, and the amount of water required to design and manufacture concrete with this methodology is based on the water content being limited to getting the concrete to a workability that is described as plastic.

Getting the desired workability from this stage onward should be achieved by the addition of mid- or high-range water reducers. Following the result is that, in the majority of situations, this procedure for concrete will have the following differences to "conventional" concretes:

1. The total cost of materials will be lower.
2. The volume of cementitious material required to achieve a given compressive strength will be lower.
3. The amount of admixture used will be higher (this is offset by the reduction in cementitious material).
4. Sand contents are higher.
5. The concrete will be preferred by pumpers, placers, and finishers.
6. Changes in workability, from 50 mm through to genuine self-consolidating concrete can be achieved from the same mix design, just altering the admixture dosage.

The potential disadvantages that can come from this methodology are when

1. The sand cost is prohibitively high.
2. The sand is of very low quality.
3. There is an imbalance in the admixture and cementitious material delivered costs.

When proportioning concrete this way, it is important to follow the admixture addition recommendations. Typically, the later the water reducers or plasticisers are added, the more effective they are or the less required. This is critical when considering cost of the materials to make the concrete.

Users may be pleasantly surprised by the performance of the concretes that are proportioned using this method, the same mix design with the exception of the admixture dosage can cover a wide range of workabilities. In some cases, designing mixes this way has allowed some "conventional" concrete designs to behave as high-quality self-compacting concretes just with the addition of extra admixture, a real bonus when aggregates or material storage or bin numbers is an issue at the plants.

8.6 INTRODUCTION OF SUPPLEMENTARY CEMENTITIOUS MATERIALS

Using this empirical mix design methodology, there are two paths that can be taken when adding supplementary cementitious materials. With current batch records, it is possible to ascertain how the materials will behave in most situations. So, when the production and quality control data are available for a set of materials that will provide a good guide as to how the components will perform in different dosages and blend percentages.

The European standard on the concrete (EN 206) sets conditions for consideration of mineral additions in partial substitution of cement in concrete in the formulation. It defines two concepts: the concept of the coefficient k and the concept of equivalent performance. Regarding the first, this standard defines a coefficient, k, taking into account the activity of mineral additives. To quantify the activity within the meaning of resistance, by transposition, we associate the resistance of mortar to those of concrete. The idea is to compare, in a given period, the compressive strength of two mortars with the same proportions of sand and water, first with cement alone, and then when in proportion of cement is replaced by a mass of mineral addition is determined experimentally by an activity index, i:

$$i = \frac{f_A}{f_{w-A}}$$

where f_{w-A} and f_A represent the strength of the mortar, either with a ratio "x" of additions substituted for cement (f_A) or without addition (f_{w-A}).

With regard to the concept of equivalent performance, the European standard states that concrete with added mineral must have similar performance to that of a reference concrete without addition, especially toward the aggression of the environment and its durability. It also provides recommendations for limits on the composition and properties of concrete according to the exposure class. The method applied involves the concept of equivalent binder. We define the equivalent amount of binder as the sum of a quantity of cement and an additional amount of the weighted coefficient k.

$$L_{eq} = C + kA$$

This k coefficient is determined for fly ash and silica fume with all types of cement. In case the cement used was CEM I type, the coefficient values are determined taking into account the specific standards.

The calculation of the coefficient k is as follows:

- It is recognised that the relationship between k and i is linear.

- It considers two special cases.
  - One where the addition is the cement itself.
  - One where the addition is absolutely inert.

If the addition is the cement itself, we have by definition of i and k: $i = 1; k = 1$.

If the addition is inert, we have by definition of k: $k = 0$ and i takes a value i_{w-A} that we need to calculate using Bolomey equation.

$f_{W-A} = \alpha\left(\dfrac{C}{W} - 0,5\right)$ and then the calculation of f_A, which is like f_{w-A} noting that the addition being totally inert, the effect of the substitution is only to reduce the cement content from C to $(1 - x)$:

$$f_A = \alpha\left[(1-x)\dfrac{C}{W} - 0,5\right]$$

That gives for inert addition

$$i_{W-A} = \dfrac{f_A}{f_{W-A}} = \dfrac{(1-x)\dfrac{W}{C} - 0,5}{\dfrac{C}{W} - 0,5}$$

In applying the formula to the following example for w/c = 0.55 and $x = 0.20$, we can calculate the value of the activity index.

$$i_{W-A} = 0,67$$

Let us now determine the linear relationship between k and i from the two following conditions:

For $i = i_{W-A} \rightarrow k = 0$

For $i = 1; \rightarrow k = 1$

$$k = 1 - \dfrac{1-i}{1-i_{W-A}}$$

By replacing i_{W-A} by its value:

$$k = 1 - \left[\dfrac{1}{x}\left(1 - 0,5\dfrac{W}{C}\right)(1-i)\right]$$

This relationship determines the value of k at any date, for a given value of substitution, and a value of w/c. It is very important to take account of changes resulting from the concrete formulation. The value of k can be greatly reduced when the w/c increases from 0.5 to 0.6.

Example 8.1: Application 1

Calculate of the resistance target at 28 days for plastic concrete dosed at 300 kg/m³ of CPA-CEM I 42.5, 66 kg/m³ fly ash (characteristic values of the activity index above 0.82 at 28 days and 0.93 at 90 days) and that have natural aggregates with Dmax equal to 20 mm.

Solution

For activity index above 0.82 at 28 days and 0.93 at 90 days, the value of k given by the standards is equal to 0.5.

C + kA = 300 + 0.5 × 66 = 333 kg/m³

Bolomey equation

E + V = 210 l

FM = 55 MPa

kb = 0.55

fc28 = 55 × 0.55 (333/210 – 0.5) = 33 MPa

Example 8.2: Application 2

Design a C25-30, workability S4 and natural aggregates (D = 20 mm), superplasticiser, and fly ash. In order to focus the calculation on cement and fly ash dosage, we are going to consider the following data:

- Water content = 185 l (we consider that the superplasticiser reduces the water demand by 12%)
- Choice of cement = CPA-CEM 42.5
- w/c = 0.6

1. Calculate cement without addition.

$$C = \frac{185}{0.6} = 308 \text{ kg/m}^3$$

2. Calculate cement and fly ash dosage in the context of substitution of cement.

Given that the cement is a 42.5, the coefficient k for fly ash is equal 0.4. Hence, a first equation:

$$C + 0.4\ FA = 308 \text{ kg/m}^3$$

The second equation corresponds to the optimal dosage in microfines. For D = 20 mm, the optimal microfines volume is 135 l/m³. Hence, we can obtain the second equation:

$$\frac{C}{3,15} + \frac{FA}{2,2} = 135$$

The solution of these two equations gives C = 265 kg/m³ and FA = 114 kg/m³.

It remains to be seen whether the ratio $FA/(C + FA)$ is less than 0.33. In fact, it is slightly higher. For this reason, the calculation must be repeated by replacing the second equation by the ratio specified by the standard, namely,

$$C + 0.4\ FA = 308 \text{ kg/m}^3$$

$$FA/(C + FA) = 0.33$$

The solution of these two equations gives C = 257 kg/m³ and FA = 127 kg/m³.

Chapter 9

Statistical analysis

Statistical analysis is not an exact science. However rigorous and elaborate the statistical techniques used, the conclusions can be no more reliable than the assumptions on which they are based. Where a limited amount of data has been obtained from a one-off experiment or series of observations, it can pay handsome dividends to apply very elaborate analysis techniques to squeeze out the last drop of knowledge. However, quality control (QC) is not a one-off experiment but a continuing flow of data. Furthermore it is a field that is, or should be, rigidly governed by economic considerations.

The requirement is to ensure a given minimum quality of concrete in the structure. This can be accomplished by using a higher average quality, at a higher cost in materials, or by achieving a lower variability through higher expenditure on control. The higher control expenditure itself can be in the form of a large amount of rough testing with little analysis or in a smaller amount of more carefully monitored testing and a more thorough analysis of the results. A balance should be sought that yields the minimum overall cost for a given required quality. The balance must take into account the standard of personnel and equipment economically available. There is no merit in devising a system that requires that every testing officer be a qualified engineer and every team include a professional statistician, if the result is a higher cost for a given minimum quality.

The concern should not be to apply elegant or rigorous statistics but only to achieve accurate control of concrete quality. Relatively crude statistical techniques can be used if their limitations are very clearly understood and the controller must always be prepared to overrule or revise unrealistic conclusions produced mathematically. It is quite difficult to do this without permitting bias to cloud judgment, but there are several factors that save it from being almost impossible. One of these is that in QC work a conclusion is usually provisional and subject to revision as further results are received, thus a downturn in results may be dismissed as a chance variation or testing error when first spotted, but if it is confirmed by subsequent results, it must then be accepted. Another is that related variables such as slump, density, and concrete temperature can confirm or deny an unusual result by indicating what caused it. Thus if a single low test result is from the lighter of a pair of specimens, it can be neglected, but if a low pair of strengths are accompanied by a high slump reading they must be accepted as fact but still may not indicate a need for a mix revision, only for better slump control.

Some crude statistical techniques have been used by the authors. This has been done quite deliberately since, in our opinion, more mathematical sophistication would not help. Rather, what is needed by way of sophistication is a very thorough realisation of what factors may cause conclusions to be unrealistic, how unrealistic they might be and what can be done to ensure that such conclusions are weeded out and do not lead to inappropriate control action. The total amount of sophistication in a scheme must be limited to keep it within the capability of ordinary practitioners. It must always be borne in mind that the objective is to achieve more economical operation rather than to display virtuosity.

9.1 NORMAL DISTRIBUTION

If a mathematical description or pattern of a set of results can be found, it may be possible to establish what the pattern is from a limited number of results already obtained and use it to predict what future results will be obtained if the current pattern continues to apply. For example, it may be possible without ever having obtained a result below some particular value to predict that a result below that value will inevitably occur unless action is taken to change the pattern. We shall be in a much stronger position to control concrete quality if it can be established that control action is necessary without experiencing even one "failure" than if we have to wait for failures before reacting to them. The position will be even stronger if it can be established from early-age tests or even from tests on the freshly supplied concrete, rather than from 28-day results.

If each result is considered as a ball and a number of slots corresponding to strength ranges are set up (e.g., 22.5 to 25 MPa, 25 to 27.5 MPa, 27.5 to 30 MPa) each result can be placed in its slot giving a picture like Figure 9.1. Such a figure is known as a histogram. If we have a very large number of balls and divide them into narrower slots, the result may approximate to the typical smooth distribution curve.

One purpose of introducing Figure 9.1 was to make it clear that area under the normal distribution curve represents number of results. Just as each ball occupies the same area in the two-dimensional representation, so each unit of area in the normal distribution represents a fixed proportion of test results. This type of graphical representation is called a frequency distribution or just a distribution. There are many different shapes of distribution curves known to statisticians but the particular bell shaped curve shown is called a normal distribution. It can be constructed from a standard

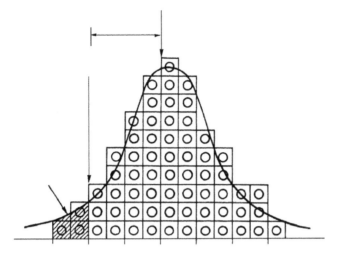

Figure 9.1 Simulated distribution of test results.

Table 9.1 Percentage of results
outside statistical limits

A (%)	k
0.1	3.09
1.0	2.33
2.5	1.96
5.0	1.65
10	1.28

table of figures (ordinates) appearing in any statistics textbook. This table will be accompanied by a second table (Table 9.1) listing the areas under the graph more than a given distance away from the mean (the high point).

The information needed to construct the graph (apart from the table of figures) is only the mean (average) of all the results, which we shall call X, and a quantity called σ or SD, which is the standard deviation and is a measure of how widely the results are spread. The numbers X and σ can be read from many simple calculators when a series of results are entered; a computer can also automatically produce them. The standard deviation is the square root of the average of the squares of all the differences between each individual result and the average of all results, that is,

$$\sigma = \sqrt{[\Sigma (x_i - x_m)^2/n]}$$

where
x_i = Individual result

x_m = Mean of all results

n = Number of results

Another method of determining the standard deviation is from the difference between successive results: Average difference/1.13. This method gives the same answer as the earlier "standard method" if the data analysed is a true normal distribution. However, there is a very useful significant difference if the data analysed is a time sequence of results having a change of mean somewhere in the sequence. In such a situation the standard method gives an inflated value for the standard deviation because it effectively involves a change of the true mean of the results both before and after the change to a new intermediate mean. We do not need to go into the mathematics of this (although they are quite simple); it is sufficient to realise that it occurs and to take it into account. The difference method is almost totally unaffected by such a change. It is particularly useful in assessing the variability of multigrade results since it is quite easy for the computer to be programmed to average differences from the last result in the same grade. In this way a much more meaningful SD can be obtained from a relatively small number of results scattered over a large number of grades.

The UK QSRMC quality control system uses the difference method since it assumes that change points will be relatively rare and effectively restarts the analysis after one has been experienced. Ken Day's QC system prints out the SD from the difference method at the top of its result table display, but then gives the SD by the standard method for each separate grade of concrete in the table itself. Of course grades with few results are likely to show large fluctuations in SD, but looking at grades with say 20 or more results, a standard method SD much in excess of the difference method SD at the top of the table usually indicates that there has been a change point in that grade, which should be investigated. However, it could also indicate that there are particular problems causing high variability in that grade (also requiring investigation).

The difference method SD can also be applied to the within-sample (or testing error) SD. Where pairs of specimens are tested, the within-sample SD is given by Average pair difference/1.13. Where three specimens are tested at the same age, the SD is given by Average range (difference between highest and lowest)/1.69.

Returning now to illustrating the principles of statistics, Figure 9.2 shows three distributions with the same standard deviation but different mean values. Figure 9.3 shows three distributions with the same mean but different values of standard deviation.

We are interested in the percentage of results less than a certain strength (i.e., the percentage defective). Looking again at Figure 9.1, the distance below the mean (or above, the curve is symmetrical) can be expressed as a parameter k (i.e., a variable number) times σ and the area as a percentage

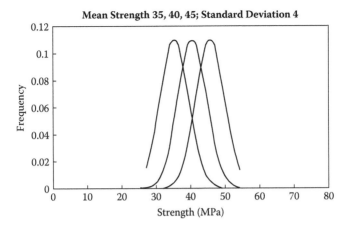

Figure 9.2 Three distributions with the same σ and different values of X.

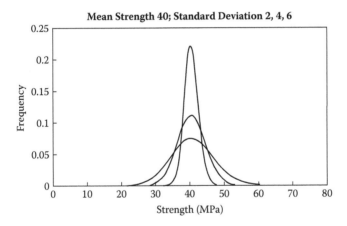

Figure 9.3 Three distributions with the same X and different values of σ.

of all results. The published tables relate the area to the value of k. Table 9.1 is an extract from such a table.

9.1.1 Permissible percentage defective

There is logic in using a 5% defective level (or even a 10% defective level) in that adherence to the assumed statistical distribution is not exact. The assumption predicts reasonably well the level below which 5% of results fall (in the authors' experience there are likely to be actually 2% to 3% below the level below which 5% are predicted to fall, but more about this later) but at the 0.1% level, the assumption has become highly theoretical and any result actually below this level is almost certainly the result of some ascertainable special cause rather than normal variability. So if the intention is to actually predict what results will be obtained, the 5% level is as far as it is reasonable to go and the use of 10% in the United States may be even more realistic. However if the results are to be judged by analysis of an adequate number of them rather than by whether any results are actually below a particular level, the situation in Figure 9.4 can be considered because it then becomes a matter not of whether the distribution is accurately followed, but simply of how much incentive it is desired to provide to achieve low variability. Figure 9.4 illustrates the available options:

Figure 9.4a shows 5% below specified strength, as used in most parts of the world.

Figure 9.4b shows the effect of decreasing the permitted percentage defective to 0.1%. This option would provide a greater financial incentive to achieve low variability (i.e., good control) but would substantially increase the average cost of concrete.

Figure 9.4c shows that by adjusting the specified strength level the average cost of concrete can be kept unchanged while still providing an increased incentive to good control.

Any suggestion to specify a 0.1% defective level is certain to encounter the criticism that this is highly theoretical and unrealistic. It is very important to clearly make the point that this is true but immaterial. What matters is to realise that it is possible to make use of any desired relative value of mean strength and standard deviation without affecting the cost of concrete from an average producer. If s is the standard deviation considered to be average, then the required mean strength (x) for a specified characteristic strength ($F'c$) could be required to be

$$x = F'c + k\sigma - (k - 1.65)s$$

or, in the United States,

$$x = F'c + k\sigma - (k - 1.28)s$$

k can be given any desired value without affecting the mean strength required of an average producer. The larger the value of k, the greater the cost advantage given to a lower variability producer and the greater the disadvantage suffered by a higher variability producer. There is no requirement to select a value of k that represents a particular percentage of results (e.g., from Table 9.1). Users should not forget the table and its significance but it may be reasonable to select a value of 1, 1.5, 2, 2.5, or 3 (or even 4, which would have no statistical significance) according to the relative importance attached to mean strength and variability.

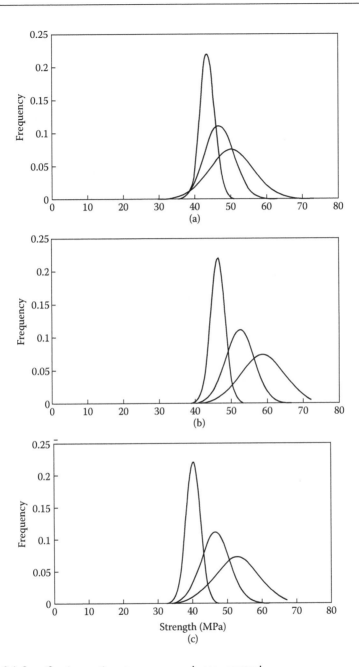

Figure 9.4 Specification options to encourage better control.

Looked at in this way, the American choice of 1.28 is seen to provide less incentive to achieve low variability than the more usual 1.64 or 1.65, and the authors would prefer to use a value of 2 or even 3. The reduced incentive may explain a reduced interest and attainment in the United States in matters of QC.

Having discounted the realism or otherwise of the theoretical percentage defective as a basis for choosing the value of k, there is another consideration. This is the accuracy with which σ can be assessed. Section 9.4 provides details.

Taking the data from Tables 9.2 and 9.3 together, it is seen that the error of estimation of the mean of three results is about five times the error in estimating the standard deviation from the last 30 results and almost four times that from 20 results. A proposal to multiply the standard deviation by 2 or 3 would therefore be reasonable if the σ were based on at least the last 30 results. However, it should be realised that a standard deviation change of less than ±25% from its previous value would not be significant.

There is a further consideration in increasing the number of results on which the standard deviation is based. If the results analysed extend across a change point in mean strength, the standard deviation will be artificially inflated. Care is necessary in determining the desired result. As discussed in Chapter 8, the variability between change points is the basic variability of

Table 9.2 Error in mean for various values of standard deviation

Standard deviation (SD) values	2	3	4
Number of results			
1	3.30	4.95	6.60
2	2.33	3.49	4.65
3	1.91	2.86	3.81
5	1.47	2.21	2.95
10	1.04	1.56	2.08
20	0.73	1.10	1.46
30	0.60	0.90	1.20

Table 9.3 Error in standard deviation for various values of true standard deviation

Standard deviation values	2	3	4
Number of results			
2	1.65	2.48	3.30
5	1.05	1.58	2.09
10	0.74	1.11	1.48
30	0.42	0.63	0.85

the production process. The frequency, extent, and time to react to change points depend largely on the control system, including control of incoming materials. The purchaser of the concrete will be interested in the overall combined effect of all causes of variability. However, a consideration of the worst concrete supplied would more accurately concentrate on the mean strength and basic variability between the two change points enclosing the concrete in question.

9.2 VARIABILITY OF MEANS OF GROUPS

So far we have considered only how well the assumption of normal distribution portrays the actual distribution of strength in the whole of the concrete. It is now time to consider how well an analysis of a limited number of samples portrays the distribution that would be obtained if the whole of the concrete supplied were made into test specimens and tested. It is conventional to consider that about 30 results are needed to give a reasonably accurate picture, but it is instructive to look into the actual situation. One way of doing this is by the use of another distribution called the Student's t distribution. This is a very useful method for evaluating comparative laboratory trials of such things as alternative admixtures or alternative cements, but it will not be considered here.

If the whole of the concrete were made into test specimens and divided into groups each of n samples, the mean of each such group would in general differ at least a little from the mean of the other groups and from the grand mean of all samples. In fact the means of the groups would be found to themselves be normally distributed but of course not so widely as the individual results. Statistical theory tells us that the standard deviation of the means of groups of n results is related to that of the individual results by the formula

$$\sigma \text{ (groups)} = \sigma \text{ (individual)}/\sqrt{n}$$

So the means of groups of 4 results will have half the σ of individual results and the mean of groups of 25 will have one-fifth the individual σ.

If we take limits within which 90% of results fall (i.e., 5% percent outside each limit) the mean of the group of n results will be within $\pm 1.65/\sqrt{n}$ of the true value. Table 9.2 summarises this.

At this point it is perhaps necessary to point out that the conformance of practice to theory is nowhere near good enough to justify the use of a second decimal place in Table 9.2. The object of the exercise is to get a feel for the order of magnitude of the errors involved.

It is worth noting that the variability of the results being examined has a strong influence on the accuracy with which they can be assessed.

This is a generally applicable statement and is another reason for preferring low variability concrete.

It will be seen that if a single test result is obtained to represent a truck of concrete, or even the mean of a pair, the assessment will not necessarily be very precise, particularly if we are dealing with variable concrete. However, variation within a batch, that is, within a single truckload, is a different matter to variability between batches and is largely a matter of testing error rather than variability of concrete (see Chapter 7). Likewise if a day's supply of concrete is assessed on the basis of three samples of concrete, a considerable error may be involved.

9.3 VARIABILITY OF STANDARD DEVIATION ASSESSMENT

In a similar manner, the value of the standard deviation (σ) obtained from analysing a limited number of results will differ from the true value for all the concrete. In this case the standard deviation of the distribution of standard deviations (no, it isn't a misprint!) is given by SD where

$$SD = \sigma/\sqrt{2n}$$

A table (Table 9.3) similar to Table 9.2 can be constructed. Although these errors are a little smaller than those in the case of the mean, they are a very much larger percentage error. Note that a group of 5 will only yield a σ value to approximately ±50% accuracy. What this means is that the variability of a group of less than 10 results simply cannot be determined with reasonable accuracy.

This has had a profound influence on the basis of specifications, because if we persist in trying to judge the quality of concrete on the basis of a small number of samples, it is not possible to give any credit for low variability (unless this is assessed on a basis external to the group of results in question). Even the inaccuracy in the mean value noted previously is large enough to require a large tolerance if good concrete is not to be rejected and this tolerance results in excessive leniency for poor concrete (see Figure 9.4a). However there is no objection to framing a criterion involving the mean of the last 3, 4, or 5 results and the standard deviation of the last 10, 20, or 30 results.

9.4 COMPONENTS OF VARIABILITY

One further piece of statistical theory is needed. This is how variabilities due to separate causes combine to give an overall variability. There is a famous example of a wrong assumption about this marring an otherwise excellent paper on concrete quality control (Graham and Martin, 1946). The square of the standard deviation is called the variance. Standard deviations are not additive, but variances are. This can be illustrated using the famous example in question (the standard deviations are in psi).

Source of error	Standard deviation (psi)
Cement (C)	240
Batching (B)	462
Testing (T)	188

The overall error is not given by $C + B + T = 890$ but by $\sqrt{(C^2 + B^2 + T^2)} = 553$. The effect of this situation is that the contribution of all but the largest component of overall variability is reduced. Thus totally eliminating cement variability would give an overall variability of $\sqrt{(B^2 + T^2)} = 499$, a reduction of only approximately 10%. But in the famous paper, the variability of the cement was further exaggerated by including the error in testing the cement and it was reported that cement variability accounted for 48.2% of total variability. This was a very significant error because it suggested that much of the variability was outside the concrete producer's control. Thus one would be led to putting much of the control effort into cement testing, instead of where it was most needed (slump control). This is a lesson that must be learned if economical control is to be achieved. The primary (largest) cause of variability must be found and control action concentrated on it (see also Pareto's principle, Chapter 10).

Of course it is necessary to monitor subsidiary causes as well to establish which is the major cause (and to check that what was initially the major cause has not been overtaken by some other cause); however the real control effort must be correctly directed.

9.5 TESTING ERROR

It has been argued elsewhere (Chapter 11) that testing itself is a significant source of error on a typical project and that it must be monitored. Day (1981) has experienced two different testing organisations testing the same truck of concrete and getting results differing by as much as 10 MPa (1450 psi) on occasions and as much as 3 MPa (435 psi) on average over a substantial number of samples (Day, 1981).

The error in question covers all aspects of taking a representative sample and casting, curing, capping, and testing specimens. It is only possible to fully establish the magnitude of this error by taking two samples from the same truck, and this is rarely economically practicable unless serious malpractice is suspected and is to be investigated for a short period. However the within-sample error can be established providing that two (or more) specimens from the same sample of concrete are tested at the same age. Day introduced a system by which the concrete supplier's own control testing was accepted as the project control providing that he produced double sets of specimens at specified intervals and delivered them to an independent laboratory for test. This is much more economical than having an independent sampler on site and avoids the concrete supplier claiming that the independent samples have been incompetently sampled, cast, or field cured. The only remaining problem is that someone has to ensure that the selection of trucks for testing is unbiased. This system is highly recommended wherever there is any concern about the veracity of the supplier's own testing. However, the net result is often that the supplier's testing is seen to be acceptable and comparative testing discontinued.

It has been pointed out that even five specimens would not permit a meaningful direct determination of standard deviation for a single sample. However, another piece of statistical theory shows that the average difference between many pairs of specimens from different samples is related to the within-sample standard deviation by the simple equation

Within sample standard deviation = Average pair difference/1.13

(In the case of sets of three specimens the difference between highest and lowest, that is, the range, may be used in the same way, and in this case the 1.13 becomes 1.69.)

Generally there is no point in converting to standard deviation for our purposes and the average pair difference is directly monitored. The best achievable average pair difference on normal concrete is 0.5 MPa (say 75 psi) and between 0.5 and 1.0 MPa can be considered acceptable. However, the authors have encountered reputable laboratories with a pair difference consistently in excess of 1.5 MPa. The seriousness of this situation can be appreciated when it is realised that even this figure does not include sampling error and that a really top class producer can work to an overall standard deviation of concrete quality below 2.0 MPa. As discussed earlier we must not fall into the error of saying that testing is three quarters of the total variability (and remember the 1.13 factor) but nevertheless such testing is grossly unfair to the producer.

9.6 COEFFICIENT OF VARIATION

Another measure of variability is the coefficient of variation. This is the standard deviation divided by the mean strength and expressed as a percentage. The question is which of the two parameters best measures relative performance on different grades of concrete. The argument resurfaces from time to time, even though in Day's opinion general agreement that standard deviation should be used was reached in the 1950s. The authors have personally monitored thousands of test results covering 20, 25, 30, 40, and 50 MPa grades of concrete from the same plant over long periods of time. There has never been any question in Day's mind that standard deviation remains reasonably constant over the 20 to 40 MPa grades (i.e., mean strengths from 25 to 45 MPa or 3600 to 6500 psi). This opinion was formed in the early 1950s when he consistently achieved a standard deviation of less than 250 psi (1.7 MPa) on very tightly controlled factory production with a mean strength in excess of 9800 psi (6.7 MPa). This was certainly abnormal concrete produced in tiny quantities and, being of earth-dry consistency, visual water control was very easy. However if this figure is expressed as a coefficient of variation of less than 3%, it would represent a standard of uniformity impossible to achieve on concrete of normal strength, even under laboratory conditions.

This firm opinion, even allowing for the quoted high-strength experience, must be tempered by an acknowledgement that a slightly higher standard deviation is normally experienced on 50 MPa and higher grades. This appears to be largely due to the greater difficulty in achieving accurate testing, perhaps in turn due to the different mode of failure of higher strength concrete (where bond failure, or even aggregate failure, rather than matrix failure tends to be experienced). The increase in both average pair difference of specimens and overall concrete standard deviation is of the order of 0.5 to 1.0 MPa.

Since publication of the first edition interesting further evidence is on hand. The Petronas Towers in Kuala Lumpur, Malaysia, project (at that time the world's tallest building,) involved more than 40,000 cubic metres of 80 MPa grade concrete. Being under a UK type specification, this required a mean strength of approximately 100 MPa (cube, at 56 days). It can be imagined that in view of the importance of the project, the initial concrete supply was at a conservatively high mean strength of just over 110 MPa. This caused the overall standard deviation for the whole of the 632 samples tested at 56 days to be inflated to 4.7 MPa. However when things had settled later in the project, a run of 237 consecutive results gave a standard deviation of 2.8 MPa with a mean strength of 99.3 MPa.

An even lower SD value of 2.6 MPa on 80 MPa concrete for the Chateaubriand bridge is reported (de Champs and Monachon, 1992).

Set against these figures are the decisions of ACI Committees 211 (Mixture Proportioning), 214 (Evaluation of Test Results), and 363 (High Strength Concrete) to adopt coefficient of variation as the meaningful index of variability. The leading advocate of this view was Jim Cook, but of course the decision was from the committees as a whole. Day (1998) suggests that high-strength concrete offers more scope for increased variability if either the testing process or the regulating analysis system is of less than the highest standard but does not necessarily have higher variability. Cook's view is that lower coefficients of variation on high-strength concrete are obtained simply because the producer is trying harder than with his normal concrete. This contrasts with the often expressed view that a producer makes his reputation on his high-strength concrete but his profit on his low-strength concrete. For this reason, Australian concrete producers are certainly trying every bit as hard to achieve low variability on their low-strength concrete. However it may well be the case in the United States, where specifications often do not allow the producer to derive any financial benefit (i.e., any cement reduction) from the attainment of lower variability.

The authors' strong advocacy of standard deviation as the measure of compressive strength variability does not mean that the coefficient of variation is a useless parameter. Obviously the same standard deviation cannot apply to such variables as tensile or flexural strength, much less to slump or density. A 5% to 10% coefficient of variation in anything generally represents a variable under reasonable control, although, for example, a modern batch plant can achieve much better than 1% in cement batch weight (if properly maintained).

9.7 PRACTICAL SIGNIFICANCE OF STATISTICAL ANALYSIS

The most obvious point emerging from the foregoing is that it is not feasible to take a quantity of concrete small enough to be regarded as a unit for purposes of acceptance or rejection, and to represent it by a sufficient number of test results to assess its quality with reasonable accuracy. It is also economically impractical to consider physically rejecting concrete that is only slightly understrength. Since the future progress of the concrete industry depends on encouraging reduced variability, it is absolutely essential that quality be assessed on the basis of a large enough pool of results to enable not only mean strength but also variability to be accurately assessed. Since no one should consider rejecting a month's concreting because compliance testing suggests it is slightly understrength, there is simply no other way to go than cash penalties or cash incentives. (Although it is feasible for the real diehards to impose this penalty in the form of increased cement content or increased testing as noted in Chapter 6.)

The next point is that we do not wish to sit back and watch the contractor dig his financial grave for a month or so without taking any action. Even worse not taking appropriate action until the concrete becomes not just contractually unacceptable but structurally unacceptable. An eventual cash penalty may bring justice to the situation and may avoid him repeating his error, but it will not provide the quality of concrete required in the current structure. Therefore a method of closely monitoring the situation and taking early action to revert to the desired quality is very desirable. This used to mean keeping a graph known as a Shewhart QC chart, however these have been superseded by cusum control charts, such as the ConAd system.

As we have seen, a substantial error is possible in assessing the standard deviation, mean, and 5% minimum of a small group of results, so that they cannot be used with any degree of fairness to reject or penalise. Nevertheless more than 50%, perhaps as much as 70% or 80%, of such assessments are quite realistic. They are therefore very useful as a guide to the state of affairs provided they are used only as a warning that the situation should be carefully considered and not as a basis for precipitate action. Having isolated the rigid legal requirement as based on an unquestionably accurate assessment of a large quantity of results, it is then possible to informally consider a large number of factors in deciding when a small mix adjustment may be desirable. There will be scope for a small difference of opinion between concrete producer and supervisor from time to time, but the latter can afford to concede graciously and wait for the fullness of time to bring retribution if it was merited, secure in the knowledge that the quality shortfall will be minor and the retribution precise, inevitable, and indisputable.

A very interesting matter is a comparison of the standard deviations considered normal in Australia and the United Kingdom. Day has for many years considered 3 MPa (say 450 psi) to be a normal figure for an average ready-mix plant in Melbourne. Of recent years the better practitioners are attaining 2 MPa or even fractionally less. In the United Kingdom, a figure of 4 to 6 MPa is considered normal. It is not likely that physical control of production is genuinely twice as good in Australia and an explanation is likely in the statistical concepts applied. In the United Kingdom, results are corrected or normalised according to cement content so as to provide a basis for combining results from different grades. It would appear that this does not work very well. Having created an artificially higher variability in this or some other manner, the task of detecting change becomes more difficult. When a rigid mathematical requirement (in the form of a V mask) is applied to determine whether an adjustment should be made, the difficulty is compounded. When adjustment is delayed in this manner, a genuinely higher variability is created or allowed to continue. This question is further examined in Chapter 11.

Chapter 10

Quality control

The purpose of quality control (QC) is to ensure the continuous production of an item of the required quality at minimum cost. Many countries still attempt to do this for concrete by testing specimens at an age of 28 days and provisionally rejecting concrete that does not comply. It is rarely feasible to actually discard concrete that has been in place for 28 days so further testing by drilling cores or by ultrasonic testing ensues. Also the concrete in trucks not sampled for testing remains under suspicion.

What is needed, after setting up appropriate production facilities and selecting suitable materials and mix designs, is to detect and rectify any departure from the intended quality at the earliest possible stage.

Although the most significant requirement of concrete may not be strength, strength is the best means of detecting change from an initially satisfactory mix. However variations in strength itself can themselves often be predicted from earlier data including slump, temperature, and especially test specimen density (which should always be determined on arrival at the laboratory rather than at test, preferably within 24 hours).

MMCQC (multigrade, multivariable, cusum QC) remains the best way of detecting change in the concrete being produced and it remains more effective to aim QC at the detection and cause of change rather than at checking conformance to a specified limit. Putting it simply, if a change is detected in the quality of the concrete being produced and supplied and the cause of that change can be established, then the change is genuine without waiting for statistical confirmation. This is the purpose of multigrade cusum graphs. For example, if strength shows a reduction from its previous average value and density also shows a reduction, while either slump or temperature (of the concrete, at the time of the slump test) shows an increase, then it is clear that the strength reduction is due to extra water. In fact it may not even be necessary to await the early strength result as it can clearly be predicted.

Of course typical concrete will no longer be composed solely of aggregates, ordinary Portland concrete (OPC), and water, so that many other causes of strength reduction are possible. Any variable (such as batch

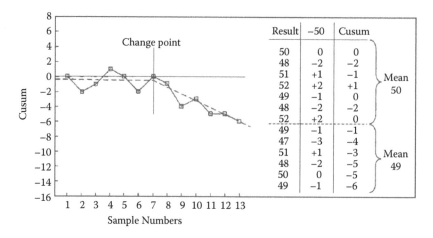

Figure 10.1 Simple cusum control chart.

quantity or quality of cement, admixture, or pozzolans) that is frequently measured can be included as a cusum and correlation sought. However, a different problem location technique is also valuable. It is important that if different types of mix are in production they are assigned to groups having one feature in common (one mix can be in several groups). The cusum program needs to be able to switch to a different group at the press of a button. So, when a downturn (or upturn) is seen on the multigrade cusum, it is easy to check which groups show the change and which do not; the cause of the change may then be obvious.

But what is a cusum graph? A cusum is the cumulative sum of differences from a target value. Figure 10.1 shows the calculation and graphing of a number of "results" varying between 45 and 55. Looking at the first column of numbers, no one could possibly see a change point in them. Graphing them as a normal direct plot would also be unlikely to reveal a change. However, subtracting 50 from each number (as an approximate mean), calculating a cumulative sum of the resulting differences, then graphing these cumulative sums shows a very clear change point in them. This is cusum analysis.

10.1 MULTIVARIABLE CUSUM

Then the question arises as to how many results after such a downturn (or upturn) are needed to provide a reasonable certainty that the change is genuine and not a temporary statistical aberration. People at QSRMC (the UK quality scheme for RMC) have calculated a series of V-masks (Figure 10.2) that provide a statistical assessment of the reliability of the downturn being genuine. I shall not trouble you with details of this because if the cause

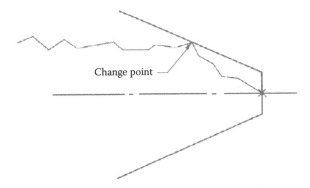

Figure 10.2 Use of V-mask on cusum chart.

of a downturn (or upturn) can be seen, then it is genuine. For example, if two or three 7-day results on a strength cusum show a downturn, and that downturn coincides with a downturn in density and an upturn in either slump or temperature, then the cause of the strength reduction is fairly clear, and the downturn is obviously genuine without waiting for a dozen or more results to provide statistical certainty.

What is required is a multivariable cusum, which is a series of cusum graphs of all the available information. The ConAd system, now owned by Command Alkon and retitled "CommandQC", provides for dozens of variables including sand grading, cement properties, admixtures, batching details, delivery truck, site sampler, and tester. Of course it does not help to have more than eight graphs on the screen at a time but they can be cycled through (but on the Pareto principle, see later).

Although a full program such as CommandQC or SpectraQEST is very worthwhile for a substantial producer, Ken Day has produced a small free program to enable smaller producers to apply the principle and also to enable specifying engineers to assure themselves, with very little time and effort, that control is effective.

It should be noted that although independent commercial laboratories can certainly provide the service (provided they make a point of distributing all results on the day they are obtained) Day's experience of over 50 years and many countries has been that the best QC is provided by a laboratory owned and operated by the concrete producer. Where this is being done and the honesty or competence of the laboratory is in question, it is generally too expensive and too difficult to organise for an independent lab to sample the same truck as the producer's team (and testing different trucks does not really prove the point). The solution is to require the truck to be sampled to be selected before its arrival by site staff (not the producer's personnel) and to require a double set of specimens to be made, with half delivered to an independent lab. The objective being to check on the producer's testing and reporting rather than the quality of his concrete. Exercises of this type are reported on Day's website, where it is also reported that the laboratory of another producer can be used as the "independent" laboratory. Where the early-age results of the two labs differ, it is easy to arrange to witness the actual 28-day tests.

10.2 MULTIGRADE CUSUM

Another concept introduced by Day is that of a multigrade cusum. Any target can be chosen for a cusum, but if the target chosen is the continuously updated mean of the item being cusummed, this places the emphasis on the detection of change rather than adherence to a target. It is perhaps not obvious that it is reasonable to combine a strength divergence from the current mean of a 20 MPa mix with that from a 100 MPa mix, but Day tried it 30-plus years ago and it works. This concept goes far beyond the "families" concept of EN206 (and is much easier to apply) and enables all test data on every mix in use (including lightweight aggregate [LWA] and normal concrete, self-compacting, and no slump) to appear on the same cusum graph. So long as the cusum graphs remain essentially horizontal, all concrete is under control. When a change point is seen, the other cusums are examined to see if they provide an explanation. If they do not, then a further concept is implemented, that of "groups".

10.2.1 Groups

The mixes in a multigrade cusum need to be collected into groups with something in common and not shared by other grades of concrete. One grade can be a member of several groups. It may be that some use a different sand, a different cement replacement material, or a different admixture. It could also be that some are only supplied to a distant site or to a particular pump known to give trouble from time to time (Pareto). The analysis program must make it easy to call up each group in turn; even the simple free program does this. The technique when a change is seen without a clear explanation is to try selecting only the results from particular groups to see which are affected and which not. This technique is particularly useful with the free program KensQC, which has only a very limited range of variables in the basic system.

10.3 MONTHLY PRINTOUT

At the end of each month, or at any other selected time, even free programs print out an analysis of all the mixes in use. Desirably this should be in order not of grade number but of departure from the target strength.

The program will be dealing with both actual 28-day results and those predicted from the 7-day results. When there is an inadequate number of results for the month in a particular grade to provide a reasonable standard deviation (SD), the program will go back to include some results from the previous month. It is important to check that the predictions from 7 days are reasonably accurate. Many technologists assume that the 7-day result will be some percentage of the 28-day result. This assumes that when a lower than usual 7-day result is obtained, it will also experience a smaller gain. Our experience is that this is completely wrong and the lower result often experiences a larger than average gain (because the 7-day result was often lower than merited owing to bad testing or curing). However a lower early result will on average lead to a lower 28-day result. To emphasise this point, using your own test data, the program produces a graph of 7-day result against 7 to 28 gain. Of course this facility can only be applied to one grade at a time.

Most codes of practice provide a table relating SD to grade strength or even assume that coefficient of variation (assuming SD increases pro-portionate to mean strength) is the best measure. A monthly tabulation (such as shown in Figure 10.3) will usually show that this is not the case, with similar SDs for all grades up to and including 40 MPa and only small increases for 50 MPa and above. However at 50 MPa and above there is a tendency for testing error to increase in even slightly substandard labs.

Major independent QC programs following the original ConAd, such as CommandQC and SpectraQEST, store every piece of data produced and can retrieve it almost instantaneously according to many selection criteria.

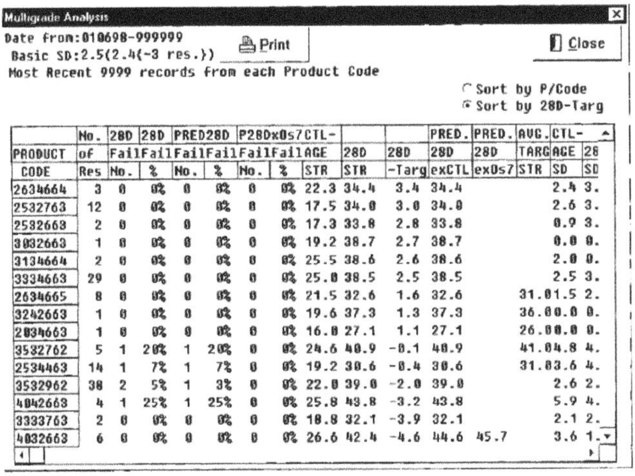

Multigrade Analysis

Date from:010698-999999
Basic SD:2.5(2.4(-3 res.))   🖶 Print   🚪 Close
Most Recent 9999 records from each Product Code

○ Sort by P/Code
● Sort by 28D-Targ

PRODUCT CODE	No. of Res	28D Fail No.	28D Fail %	PRED28D Fail No.	PRED28D Fail %	P28DxOs7 Fail No.	CTL- Fail %	AGE STR	28D STR	28D -Targ	PRED. 28D exCTL	PRED. 28D exOs7	AVG. TARG STR SD	CTL- AGE 28 SD
2634664	3	0	0%	0	0%	0	0%	22.3	34.4	3.4	34.4		2.4	3.
2532763	12	0	0%	0	0%	0	0%	17.5	34.0	3.0	34.0		2.6	3.
2532663	2	0	0%	0	0%	0	0%	17.3	33.8	2.8	33.8		0.9	3.
3032663	1	0	0%	0	0%	0	0%	19.2	38.7	2.7	38.7		0.0	0.
3134664	2	0	0%	0	0%	0	0%	25.5	38.6	2.6	38.6		2.0	0.
3334663	29	0	0%	0	0%	0	0%	25.0	38.5	2.5	38.5		2.5	3.
2634665	8	0	0%	0	0%	0	0%	21.5	32.6	1.6	32.6	31.0	1.5	2.
3242663	1	0	0%	0	0%	0	0%	19.6	37.3	1.3	37.3	36.0	0.0	0.
2034663	1	0	0%	0	0%	0	0%	16.0	27.1	1.1	27.1	26.0	0.0	0.
3532762	5	1	20%	1	20%	0	0%	24.6	40.9	-0.1	40.9	41.0	4.8	4.
2534463	14	1	7%	1	7%	0	0%	19.2	30.6	-0.4	30.6	31.0	3.6	4.
3532962	38	2	5%	1	3%	0	0%	22.0	39.0	-2.0	39.0		2.6	2.
4042663	4	1	25%	1	25%	0	0%	25.8	43.8	-3.2	43.8		5.9	4.
3333763	2	0	0%	0	0%	0	0%	18.8	32.1	-3.9	32.1		2.1	2.
4032663	6	0	0%	0	0%	0	0%	26.6	42.4	-4.6	44.6	45.7	3.6	1.

Figure 10.3 Monthly (or other period) printout.

10.4 DISTRIBUTION PATTERN

Most investigators agree that strength is at least approximately a normally distributed variable. This means that it can be completely described by a mean strength and a standard deviation, that is, the percentage of results lying above or below any particular value can be calculated from the mean strength, the standard deviation, and a table of values from a statistical textbook as seen in Figure 10.4. The authors have found this assumption to be well justified in practice except that only about half the results theoretically expected to be below the mean minus 1.64 σ usually occur in practice.

The formula used is $X = F + k\sigma$, where X is the required average strength, F the specified strength, σ the standard deviation, and k is a constant depending on the proportion of results permitted to be below F. In fact the overall distribution is likely to consist of a number of subdistributions, each with a slightly different mean strength but probably with a similar SD slightly lower than the overall SD as shown in Figure 10.5. The assumption is that there is a basic variability in the process with hopefully infrequent occurrences of an unusual factor.

The SD can be calculated in at least two different ways. The "traditional" way (by which it is defined, and which the typical calculator uses) is by calculating the mean, totaling the squared differences of each individual result from it, and then finding the square root of that total.

A second way (referred to herein as the "basic" method) is to average the difference between successive results and divide by 1.13. If (as assumed) the mean has continued to be the same during the entire string of results,

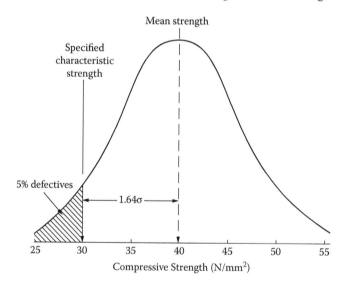

Figure 10.4 The normal distribution.

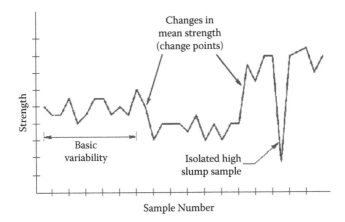

Figure 10.5 Change points and basic variability.

the two methods of calculation should give the same result, but if there have been one or more changes in mean strength, as envisaged earlier, the traditional way of calculating SD will give an increased value but the basic method will be largely unaffected. This is valuable when assessing the situation with a monthly printout table of what may be a large number of grades of concrete, since it directs attention immediately to any grades that have experienced a change during the month (or other period). Of course such a change should have been detected from early age results when it occurred and quickly rectified.

It is likely that the basic variability is due to such items as batching accuracy, moisture/slump regulation, waiting times, and so forth, whereas the "infrequent occurrences" may relate to changes in fine aggregate grading, cement or admixture properties, a failure to allow for seasonal temperature change, and so on.

In a well-run plant with low basic variability, the QC system may be primarily aimed at the early detection and identification of the infrequent occurrences and the strong advocation of cusum in this chapter has this objective. However arriving at the condition of low basic variability is obviously also important and will, in addition to cusum graphing, involve truck-to-truck variation better revealed by direct plots.

The purpose of a QC system is to ensure the provision of satisfactory concrete at a minimum cost. To do this it needs to be able to detect any change in the quality of concrete being produced at the earliest possible moment commensurate with realistic expense and time frame.

This involves a judgment of the relative cost of providing a higher margin between mean and specified strengths (or other criteria such as permeability affecting long-term durability) or spending more on control.

The cost of a higher margin is easily calculated: Cost of additional mean strength = Specified strength + 1.65 × SD. With SDs ranging from less than 2 MPa to more than 6 MPa, the producer with poor control has effectively to supply the next higher grade of concrete at the price of a well-controlled lower grade.

The cost of better control is not so easily calculated as it is strongly affected by both the choice of control system and the degree of understanding of several basic concepts. Some of these concepts are

1. Professor J.M. Juran's dictum that QC of any manufactured item should be aimed at "controlling the mass and not the piece". Thus control should be aimed at controlling the whole production of a plant or even a group of plants rather than individual batches or isolated deliveries to a particular site of a particular grade or on a particular day.
2. Pareto's principle of locating the main causes of variability for particular circumstances and concentrating control on them, rather than spreading it more thinly over all possible causes. (Pareto was an Italian economist engaged in trying to discover the sources of Italy's wealth. He found that in any town, more than half of the total wealth was under the control of four or five men and he could get a better answer by first finding these men and then asking his questions rather than attempting a general survey of the population.)
3. The purpose of testing anything is not to discover and reject unsatisfactory items but to establish the minimum quality level of the whole. (If one in ten deliveries is tested, then for every individual item rejected, nine equally defective items will have been accepted.)
4. There is a difference between unsatisfactory and unacceptable concrete. Unsatisfactory concrete (that which does not quite meet the specification but is not dangerously defective) can be detected and financially penalised on the basis of a statistical analysis of a month's test results without further investigation. It has to be recognised that, under perfect control, there is 1 chance in 1,000 of a result 3.09 SD below the mean. More realistically this can be seen as only 1 chance in 1000 that such a result is a consequence of normal variability rather than a particular problem. Again, more realistically, there is 1 chance in 100 of a result 2.33 SD below the mean and this would be 1 × SD below the specified strength, so such a result cannot be considered to be unacceptable.

   Any unacceptable concrete has to lead to extensive in situ investigation of all concrete of that grade during the period and would constitute a huge problem for both the producer and the site team. (Experience is that if unsatisfactory concrete is always detected and penalised, dangerously defective concrete does not occur, but if unsatisfactory concrete is supplied with impunity, dangerously defective concrete is likely to follow.)

It is obvious that these principles, and the requirement of early detection, are not met by the current practice of statistically analysing a number of 28-day test results and following up any marginally unsatisfactory result with coring or nondestructive testing (NDT).

What is required is to examine early-age results and to follow up any predicted departure from the required mean strength with an immediate mix adjustment or other effective action to restore the required mean. It may be good practice to make slightly more than the required increase in the case of a shortfall and slightly less than the calculated saving in the case of a higher strength than necessary, that is, to ensure that any shortfall is definitely immediately remedied and to approach the required mean with caution. This kind of immediate reaction to observed early-age variation cannot be imposed by a purchaser or supervising engineer, and must be in the hands of the producer.

Another question is the extent to which mixes should be deliberately varied at the time of batching in an attempt to avoid changes in strength. For example, it is known that a higher concrete temperature at the time of batching will increase water requirement and therefore reduce strength, yet almost any set of traditional test results will show a strength reduction in early summer and a strength increase in early winter. It takes some traditional control systems more than a month to react to this situation and adjust mixes. There can be a distinct difference between concrete temperature on a cold morning and a hot afternoon and Day has written about concrete on a cold morning being rejected as of excessive slump when it might in fact have a lower water to cement (w/c) ratio and a higher eventual strength than lower slump concrete on the same afternoon. However, the whole question of whether it is ever reasonable to reject an individual truck of concrete on slump or appearance grounds needs careful consideration.

There may be genuine reasons for requiring a higher slump (or a longer delivery time) on some deliveries and rather than having water added on site the concrete could be officially supplied at a higher slump, probably increasing the admixture dosage rather than adjusting water and cementitious quantities.

Variation in cement, admixture, or aggregate properties may also be known in advance of batching in some cases. Of course this possibility of "just-in-time" mix variation would require close regulation and skilled staff if it is not to lead to even worse variability. Any such policy should be introduced on a well-scrutinised trial basis and continued only if shown to reduce test result variability.

Testing is normally one at 7 days and two at 28 days. Some operators tend to keep the second of the two 28-day results for test at 56 days if the first is below the anticipated strength. In contrast to this, our recommendation is to bring forward one of the 28-day specimens for test at 7 days if the 7-day specimen gives an unexplained low result. It is a matter of determining a course of action rather than making a second attempt to obtain an acceptable result.

Some operators may have taken the precaution at an early stage of making one or more additional specimens to establish a conversion factor to predict 7-day results from 3-day results. If so, it is an even more satisfactory solution to bring forward one or more intended subsequent 7-day results for test at 3 days to see whether an apparent shortfall is continued.

The use of 7 days as an early-age test is a consequence of many laboratories only operating 5 or 6 days per week. If 7-day laboratory operation is in force, 3 days is an equally satisfactory early-age test (although more attention is required to it being at close to 72 hours and to curing temperature having been under good control). Control can also be based on testing at 24 hours or less, but in this case it is strictly necessary to insert a thermal probe to establish the exact Arrhenius "equivalent age" as explained in Chapter 7.

There is only one effective answer to "detecting any change in the quality of the concrete at the earliest possible moment" and that is by cusum analysis. As already noted, the strength of a particular grade of concrete may not be the most important requirement, but strength is the most effective parameter for the detection of a change in quality.

10.5 NORMAL DISTRIBUTION

The assumption is normally made that the results for any individual grade will form a normal distribution. We have found this to be a reasonable assumption, except that the percent below mean −1.65 SD is often 3 or 4 rather than the anticipated 5. This may be due to control action on extreme high slump.

What is of more importance is the occurrence of double-peaked or excessively skew distributions. A double peak is usually a sign of two different distributions being combined. When this occurs an attempt should be made to separate the two sources as one may be of inadequate strength, which is masked by combination with the other.

If a distribution is skewed on the high side, it may be that low results are being withheld and this possibility should be followed up. If a distribution is skewed to the low side, it may be that the coarse aggregate has limiting either strength or bond characteristics. Alternatively it may be that the operator is afraid of explosive failure or wishing to avoid cleaning up after such a failure. Both these possibilities should be avoided by the provision of a circular shield during testing.

10.6 DATA RETRIEVAL AND ANALYSIS/ CONAD SYSTEM

10.6.1 Coping with data

A basic challenge in the quality control of concrete is to cope with the availability of possibly excessive amounts of data. There is no doubt that facts can be harder rather than easier to deduce if included in more data than a person can cope with. It should not be forgotten that quality control is an exercise in cost reduction and that cost includes the cost of the quality control. A better quality concrete can be purchased at a higher price, but the task of quality control is to deliver concrete of a chosen quality at the minimum cost.

So the value of given data should be considered alongside the cost of acquiring, storing, analysing, and employing those data. In particular no substantial cost should be incurred in acquiring and storing data that will definitely not be used. On the other hand, storage of huge amounts of data is no longer a problem, providing it can be acquired at negligible cost and effort and the precise data needed can be automatically retrieved with little effort.

An example of inadequate cost–benefit occurred in the 1980s in New York where inspectors were employed to manually write down batch quantities at substantial cost, but no analysis of the acquired data was carried out. In contrast batch quantities (intended and actual) are automatically acquired electronically by the ConAd system, are automatically matched with test data on tested loads, and errors can be automatically displayed either numerically or graphically. In the latest development, the system can automatically e-mail or telephone selected personnel to advise of errors, and can predict the strength of a miss-batched load. Long-term trends in inaccuracy can be precisely displayed graphically. Of course these facilities require both suitable batching equipment and a suitable analysis program.

Other data that can be automatically acquired include details of the original order, so that a field testing officer only needs to record a batch number and his actual measurements. Also many laboratory testing machines are able to output test results direct to a laboratory computer. This not only saves time but also avoids the possibility of error in transference and the necessity to check for such errors. Not only crushing loads but also weights and dimensions of compression specimens are often automatically recorded.

The other end of the process is the retrieval of data from storage. If large amounts of data are recorded, then retrieval must be substantially automated. The largest amount of data is usually batching data. This is required in full so that cumulative errors and the variability of the process can be studied. It is not enough to have all the information tabulated so that the analyst can run their eye down the column to look for exceptions. It is not even enough to graph the data, revealing exceptions many times more quickly. It is necessary for the system to be able to retrieve those items, and only those items, having an error in excess of any nominated amount. It is also necessary to have cumulative error graphs, showing whether consumption averages that planned.

A small free program called KensQC is provided on Day's website and described next.

10.6.2 KensQC

KensQC is the program as you see it when you download it from the website http://www.kenday.id.au and install it on your computer.

The program is not really meant to display substantial numbers of results on graphs but rather to use the cusum graphs to detect change (and its cause) at the earliest possible stage. However the tabulated data on the "Multigrade Stats" printout is useful to show change over a period.

The results currently in the program are not an ideal demonstration of the value of the program, especially since they are only of a single grade, but some points can be noted:

1. The average pair difference of 1.1 MPa in 28-day test results is a reasonable but not high standard of testing. However it does show an improvement after 22/12/2011 from an initially less satisfactory figure. The earlier period showed several pair differences of 3 to 6 MPa rather than a figure generally higher than 1.1. As is fairly normal, such pairs showed one result lower than usual rather than one result higher than usual, suggesting that the higher result of a pair is more likely to be the correct value.
2. There is very good agreement between 7-day and 28-day results, showing that change points can be reliably detected at the earlier age (see especially 27/12/2011, 28/01/2012, and 17/2/2012).
3. There is some correlation between density and strength, and some reverse correlation between temperature and density and strength, showing the effect of temperature on water demand and therefore strength.
4. It is interesting that temperature shows some correlation with 7- to 28-day strength gain, suggesting that the 7-day specimens were slightly less mature in the cooler weather (it is not known whether

curing boxes were used on site and at what age the specimens were typically collected).
5. Unusually, strength shows little reverse correlation with slump (possibly suggesting that slump tests were not well performed).

Direct plot graphs

Direct plot graphs are not as informative as cusum graphs but do show where a few obvious errors have occurred. These graphs should also be briefly examined daily and an attempt made to determine the cause and significance of any errors. The results showing high 7- to 28-day gain can be caused by bad operator performance (lab or field) but are unlikely to be a property of the concrete.

Multigrade statistics table

The "Multigrade Statistics" are in fact all of the same 32 MPa grade, covering the three months December 2011 and January and February 2012. They show:

1. A slightly inadequate strength overall, arising in January, with December and February both acceptable. An overall SD of 3.89, surprisingly slightly greater at 28 days than at 7 days. This is a situation where a very minor cash penalty might be appropriate rather than any further investigation.
2. The problem arose due to slightly increased variability rather than a lower strength. Of course a slightly higher strength margin should have been used, and the cash penalty should desirably be about double the cost of providing the higher margin, and applied to all concrete supplied in January.

Use of the program

Download the program from the website http://www.kenday.id.au and key "Export Results" at the bottom of the screen. This provides an Excel spreadsheet. Copy your own results a column at a time into the spreadsheet. Only the data columns are required, as the calculated columns will automatically recalculate, deleting existing calculated items. It is important to delete any data not replaced at the foot of the spreadsheet or if you only have one 28-day result as, if only one column were replaced, the program would average the remaining column data with your own data.

Now return to the downloaded program and key "Import Results" at the bottom of the screen. It does not matter whether your results are in strict date order or sorted into grades, as the program will attend to this.

At the top of the screen you choose sort by DATE (Grade is only provided in case you wish to extract or delete a particular grade).

Then enter Date Range. It does not matter if the range entered is too wide, but if it does not include all your results, results outside the data range will be excluded from the analysis, they will not be deleted and will reappear and be included in subsequent analyses if the date range is extended.

"Refresh" will cause a recalculation, but this is automatic in most circumstances.

"Groups" is important for multigrade data. Keying "groups" will produce a screen listing all multigrade designations on the right-hand side. On the left-hand side you can successively enter any number of group names and state which of the grades are to be in that group. A group has to have one thing in common not shared with grades outside the group, for example, a particular cement or cement replacement material, a particular fine or coarse aggregate, or a particular admixture. Any one grade can be in multiple groups.

On the second line, the program displays "Basic SD". This is a figure obtained by dividing the average difference between successive results in the same grade (regardless of how many results from other grades are between them) by 1.13. If the mean result in that grade remains unchanged, this figure will be the same as that traditionally calculated, but if there has been a change in mean, this figure will be comparatively unaffected and will therefore often be lower than the traditional value.

The program then lists the total number of records (which will be seen to change if the date range is changed to include more or fewer of the results entered and "Refresh" keyed).

Finally, in a multigroup situation, the number of results in the currently selected group is given.

Transferring now to the bottom of the screen, the graphs illustrated earlier can be viewed. Using "Select Graphs Variables" the user can usually obtain a clearer picture of the situation.

The whole concept of quality control by this program is to see graphs of other variables exhibiting the same change point as the strength graphs. If a strength change point in an early-age graph is explained by changes in other variables then it is a genuine change without waiting for 28-day results or statistical confirmation. A lower density/unit weight (which should be obtained within 24 hours), especially if combined with an increased slump or temperature, will indicate a higher water content and a lower 28-day strength can be predicted even before a 7-day strength has been obtained.

The KensQC program has a very limited range of variables, whereas the original ConAd program, now CommandQC, provided a large range of variables including batch quantity data and test data on cement and aggregates.

The Group concept, if there are several different groups, may partially remedy this situation. It is very easy to switch from one group to another in KensQC and if one group is affected and others not; the cause is often clear.

Keying "Group Monthly Stats" will provide an analysis of all the data input (providing you have a group "All" nominated) split into grades and months (providing you have ticked "Split into Months" in the bottom left-hand side corner of this screen) and you can scroll down to see at the end all the results for the whole period selected.

This screen can be exported and printed. It needs to be printed landscape and even so will need to be contracted sideways slightly to fit on an A4 sheet.

Reading the columns on this screen we have Month, Group, Grade Strength, Target (the target is the entered grade strength plus 1.65 times the SD of the last 15 results in that grade).

The next column is headed P28/S28 but it means either or, not divided by. P28 is the predicted 28-day result obtained by adding the current average 7 to 28-day strength gain to the average 7-day for the month. This is replaced by the actual average 28-day result for the month as soon as that is available.

The next column is Margin, the difference between the P28/S28 and the target. If this is negative, that is when a graduated cash penalty would be justified.

- The next four columns are numbers of results: S is the total for the month, then the number of 7s and 28s and finally the number that failed to reach the specified strength. So this highlights if any results are missing and if there were any failures during the month.
- Next comes Slump, Temperature, Air%, Density (ignore Adj), and actual SD at 7 days and 28 days. There are two ways of calculating SD: one is by averaging squared differences from the mean and the other by dividing the average difference between successive results by 1.13. If the mean remains unchanged, both methods should give approximately the same answer, but the first (the traditional method) is much more affected by a change in mean than the other (basic) method. So looking at these two numbers gives an indication of whether there has been a change point during the month.
- The final column, Perr, is the error of prediction of the 28 from the 7 for the month.

Now keying "Close" on the GMS screen, we return to the main screen. The main thing now is to examine the graphs.

The CUSUM Plot is the basis of the whole concept. It will hopefully show you a correlation between strength variations and the factors causing them in your results.

Chapter 11

Unchanging concepts

11.1 CASH PENALTY SPECIFICATION

The authors believe that a cash penalty basis can provide fully fair and effective regulation of concrete strength (and thereby, quality). However, many in the industry or its clients have been reluctant to accept it. The most pressing reason why concrete might desirably incur a penalty is in fairness to other suppliers who allowed in their quotation to supply the specified strength in full and thereby failed to obtain the contract to supply. If well-intentioned suppliers do not see this as an advantage, then so be it. However, the section remains in the book to satisfy the authors' conscience that they have done everything reasonably possible to bring about this desirable reform.

This section was first published by Day (1982b) as an article in *Concrete International: Design and Construction* in 1982 under the title "Cash Penalty Specifications Can Be Fair and Effective". Permission granted by the American Concrete Institute to reproduce it here is gratefully acknowledged.

A cash penalty of twice the cost of the extra cement that would have been required to avoid defectiveness is proposed. It is shown in detail that if this is based on the statistical analysis of any 30 consecutive 28-day test results, very little inequity would result to either party (in contrast to the substantial risk of inequity under current specifications based on inaccurate, small sample criteria). The aspect of legal enforceability is considered and examples are provided of a suitable cash penalty provision used in a major Australian structure, and of several situations where cash penalty provisions would have been desirable.

A good specification system accomplishes the following (Day, 1961):

1. Ensures the detection and penalisation of unsatisfactory concrete
2. Avoids the penalisation of good concrete
3. Encourages good quality control
4. Avoids any doubt of fairness and eliminates disputes
5. Is based on sound theoretical principles

Typical concrete specifications around the world continue to levy one penalty of rejection and continue to base judgment on criteria that are known to be inefficient at distinguishing the actual quality of the concrete assessed (Chung, 1978). The result of this ostrich-like attitude is to leave supervising engineers in untenable positions, to subject concrete suppliers to gross unfairness on occasions, frequently to allow unsatisfactory concrete to be supplied with impunity, and worst of all, to fail to encourage responsible producers of low-variability concrete.

The proposed system

The quality of concrete is assumed to be represented by the mean and standard deviation of strength. Quality should be specified by the requirement:

Any deficiency in strength can be readily assessed in terms of inadequate mean strength. The cost of remedying that deficiency can be readily assessed in terms of cement content or admixture to achieve the required water to cementitious materials (w/cm) ratio.

For a limited extent of deficiency, a penalty of twice the cost of remedying the deficiency could be imposed. This penalty is negligible for small deficiencies, but if the criterion is sufficiently accurate, the penalty will be sufficient to ensure that no concrete supplier can make additional profit by supplying understrength concrete. This penalty system benefits producers of low-variability concrete and encourages improved quality control.

The key to this system is the determination of the values of mean strength and standard deviation with sufficient accuracy, and the selection of a suitable value for k. It is immaterial whether the cement-content change required to provide a given strength change is truly a constant for all concrete, providing the change is never more than twice the assumed value.

Accuracy of assessment

The gross inaccuracy of assessment encountered under most specifications arises from an inadequate number of test results (Chung, 1978) and from attempting to assess the quality of an amount of concrete sufficiently small to accept or reject as a whole. There is no such requirement in a cash-penalty specification.

A secondary reason for basing a criterion on a small number of results is to enable a judgment to be made quickly, thus limiting the amount of defective concrete supplied before a halt is called. This pious intention may become a joke when the results are obtained at 28 days.

The solution to this dilemma is to separate the functions of (1) acceptance/penalisation and (2) detection and arrest of adverse quality.

An interesting and valuable result of operating under a cash-penalty scheme is that the interests of the supervisor and the concrete supplier coincide in their joint desire to detect and eliminate adverse trends at the earliest possible moment. This cooperative type of relationship is in contrast to the traditional requirement to establish with legal precision that concrete strength is inadequate and then require the unwilling supplier to rectify the matter.

The suppliers generally recognise that rapid reaction to warnings of low strength from the quality control engineer can save the supplier money. A graphing system can provide such information based on a few early-age test results and will enable the supplier not only to avoid extensive periods of low strength but also to reduce the overall variability (a double saving in potential penalties) (Day, 1981).

The expression gives a standard error of approximately 0.74 MPa (107 psi). This means that 90% of assessments will be within 1.22 MPa (177 psi) of the correct value.

If it is further assumed that a 1 MPa (145 psi) strength change requires 7–8 kg/m^3 (11–14 lb/yd^3) of cement change (the actual value could range from 5 to 10 kg/m3 [8–17 lb/yd^3)] for different concretes) or equivalent water reduction, then the inaccuracy amounts to a maximum of ±10 kg/m^3 (±17 lb/yd^3) in cement content, or a cost of around $1.80 (Australian)/m^3 (in U.S. dollars approximately $1.50/yd^3).

Operation of the system

The specification might then read as follows.

> The specified strength of the concrete shall be X MPa and for every 1 MPa (145 psi) that the mean strength of any 30 consecutive samples minus 1.28 times the standard deviation of strength of those samples falls below X MPa, the contractor shall pay a penalty of AU$2.70/m^3 or US$2.25/yd^3) of the whole of the concrete represented by the 30 results in question.

($1 equals twice the cost of the 7.5 kg [16.5 lb] of cement assumed to be required to increase the concrete strength by 1 MPa [145 psi]).

To avoid occasional unmerited penalties under such a specification, the concrete supplier would have to incorporate 10 kg/m^3 (17 lb/yd^3) excess cement content, increasing the cost of concrete by $1.80 (Australian)/m^3 ($1.50 (US)/yd^3) above the cost strictly required, with the idea that this increase in cost is justified by the quality control benefits of the entire system.

On the other hand, a concrete supplier would occasionally escape penalisation when actually supplying concrete as much as 1.22 MPa (177 psi) under strength. If the supplier decided not to add additional cement or admixture, on average, he would be paying a penalty of AU$3.30/m^3 US$2.75/yd^3) or 75% more than the cost of virtually eliminating the risk of a penalty.

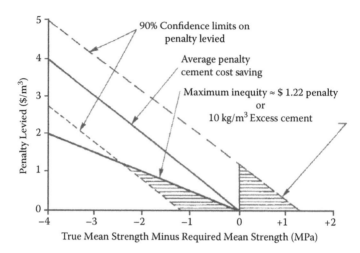

Figure 11.1 Graph of average penalty applied.

Figure 11.1 shows the average penalty that would be applied and the 90% confidence limits on that penalty for strength shortfalls up to 4 MPa (580 psi). The graph shows there is very little risk of any significant unmerited penalty and even less chance of the cement saving outweighing the penalty.

Effect of *k* value changes

The effect of an increasing *k* value would be to increase the required mean strength. This could be offset by a reduction in the specified strength below that used in the structural design. The effect of such a compensated increase in *k* value would be to provide a greater incentive to attain a low variability in the concrete strength by imposing a larger safety margin on suppliers of higher variability concrete. The actual minimum strength (say, the 3 standard deviation limit below which only one in a thousand results would fall) would be raised by such a specification.

In the authors' view, an increased incentive to reduce variability and increase security against the occurrence of very low strengths would be highly desirable. It is suggested to use a *k* value of 3 and to reduce the specified strength by 5 MPa (725 psi) in compensation.

For a *k* value of 1.28 (existing U.S. practice) and a specified strength of 30 MPa (4348 psi), the effect of this would be

2.5 MPa (362 psi) (good control)
- Required mean strength 30 + (1.28 × 2.5) = 33.2 MPa (4811 psi)
- Effective minimum strength 33.2 − (3 × 2.5) = 25.7 MPa (3725 psi)

5 MPa (725 psi) (poor control)
- Required mean strength 30 + (1.28 × 5.0) = 36.4 MPa (5275 psi)
- Effective minimum strength 36.4 − (3 × 5.0) 21.4 MPa (3101 psi)

For a k value of 3.0 (preferred), and a specified strength of 25 MPa (3623 psi), the effect would be

2.5 MPa (362 psi)
- Required mean strength 25 + (3 × 2.5) = 32.5 MPa (4710 psi)
- Effective minimum strength 32.5 − (3 × 2.5) = 25 MPa (3623 psi)

5 MPa (725 psi):
- Required mean strength 25 + (3 × 5.0) = 40 MPa (5797 psi)
- Effective minimum strength 40 − (3 x 5.0) = 25 MPa (3623 psi)

The effect of the change would be to worsen the competitive position of the high-variability supplier and limit the occurrence of occasional low strengths in the concrete supplied. The low-variability supplier would be virtually unaffected, except for the supplier's improved competitive position.

Figure 11.2 shows the relative situation under exact compliance of a 10% defective criterion for both the high- and low-variability supplier. The upper graph shows that under the present (U.S.) 10% defective basis, the low-variability supplier has a reduced incentive and the high-variability concrete includes some deliveries of very low strength. The lower graph shows an enhanced competitive position for the low-variability under the proposed 0.1% defective basis. Both suppliers in this case provide effectively the same minimum strength.

The benefits of low-variability concrete are substantial:

1. Helpful to the concrete placing crew
2. More uniform compaction
3. More uniform appearance
4. More accurately assessed on a given number of test results (possibly less frequent testing required)
5. Better control of pumping

Influence of change points

The proposed technique assumes that there will be a gradual drift of either mean strength or variability, and that it will be legitimate to select 30 results incorporating the worst period. Analysis has shown, however, that changes are usually step changes rather than gradual drifts. Thus, a specific number of results constitute the low period and all of them (and no more) should be analysed to represent the low period rather than taking an arbitrary 30 results. This is too complicated and indefinite for use in a specification

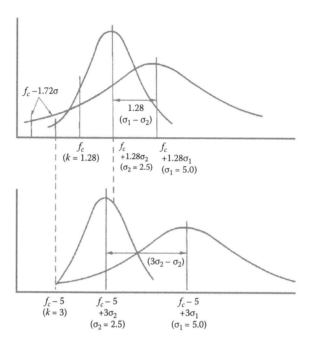

Figure 11.2 **Effect of compensated increase in k is to improve competitive position of low-variability supplier and rule out low results from high-variability supplier.**

but could be applied with mutual agreement in practice. The effect of ana-lysing 30 results overlapping a change point is to give an artificially inflated standard deviation, which is only slightly compensated for by the increased mean strength obtained from the inclusion of a few higher results and, therefore, causes a higher penalty. An alternative, slightly lower penalty based on the actual defective period can be offered, but the specification can be strictly enforced without substantial unfairness.

Figure 11.3 shows a run of understrength results that merit a penalty. Under the proposed specification, the lowest 30-result section (representing 600 m³ [785 yd³] of concrete) must form the basis.[1] A penalty of $6.16 m³ would be applied, totaling $3696.

Close analysis, however, reveals that the low strength concrete is con-fined to a 20–result section (representing 400 m³ [523 yd³] of concrete).[2] The penalty/m³ based on the 20 results would be greater but the overall penalty would be less at $7.67 × 400 = $3068. The latter penalty is the more equitable and is the one that should actually be imposed. However,

---

[1] The average strength over the thirty results in 31.56 MPa (or 2.28 MPa less than the target of 33.84 MPa based on a SD of 3 MPa and K = 1.28).
[2] The average strength over the twenty results is 31 MPa (or 2.84 MPa less than the target).

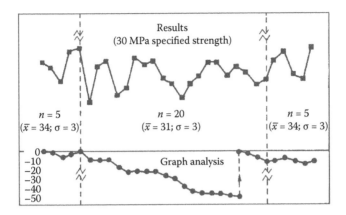

Figure 11.3 Graphical analysis of run of understrength results, which merits a penalty.

the difference is only $628 and the 30-result basis is reasonably satisfactory and much simpler to incorporate into a specification.

The assumption is that the concrete supplier would have had to spend approximately $4.00 m³ in extra cement on the 400 m³ (523 yd³) of concrete to avoid penalisation (total saving: approximately $2000 in cement cost), so the net cost to the supplier is approximately $1600. Obviously, the supplier would prefer to pay this penalty rather than delay the work and pay the costs of coring and investigating 400 m³ (523 yd³) of concrete, with the risk that some or all of it might be rejected.

Importance of quality of testing

It is obvious that the test results forming the basis for a cash penalty should provide an accurate assessment of the quality of concrete as supplied by the producer. This is by no means something that is easy or can be taken for granted. A minimum requirement is that samples should be taken, cured, and tested by a competent, accredited, and preferably independent organisation.

The best criterion of testing accuracy is the average difference of pairs of test results from the same sample of concrete. This average difference should not exceed 1 MPa (145 psi) for normal concrete (specified strength less than 50 MPa [7246 psi] and possibly excluding very low slump mixes). It is suggested that the highest of a pair of specimens is likely to be a better estimate of the true concrete strength than the mean of the pair.

The person responsible for result analysis should be alert for clearly established cases of incomplete compaction and improper curing and testing, and should be prepared to exclude such results from a penalty assessment. The previously recommended graphical analysis system, including analysis of related variables such as slump, strength, and testing, has been found valuable in distinguishing causes of variability and early detection of problems.

Parallel tests by two laboratories on the same truck of concrete reveal useful information and should be arranged from time to time.

The whole question of the reliability of concrete testing results is a matter that has received far too little attention. However, it is not a valid reason for failing to institute the type of cash penalty specification advocated here, as it causes even more trouble under existing types of specification.

No one can afford cheap testing. The best prospect of reducing testing costs is to reduce the frequency of testing, made possible by better testing, better specifications, better analysis of results, and a reduction in the variability of concrete.

Legal enforceability

Extremely crude forms of penalty are sometimes encountered, particularly on government work. Such penalties are enforced on the basis that future contracts will be withheld if they are disputed.

In British and Australian law, the key to legal acceptability is to relate the penalty to the harm suffered. It is assumed that a building owner would prefer to pay for the grade of concrete specified rather than accept a lower grade of concrete at lower cost. If the owner is supplied a lower strength concrete than specified, then he must have suffered harm in excess of the cost difference (in terms of margin of safety, durability, etc.) between the two strength levels.

Actually, the penalties considered here are too small to be worth a contractor's expense to legally challenge. However, the penalties are sufficient to ensure his cooperation in avoiding them. What the law does object to are penalties specified to scare the contractor into compliance.

Experience in Australia

Although this proposal is now 20 years old, it has been applied to only one major contract by Ken Day. This was the Victorian Arts Centre (the Melbourne equivalent of the Sydney Opera House). On only one occasion did the results actually merit a cash penalty, which was paid.

However, thousands of cubic metres of concrete have been supplied to dozens of structures using the previously discussed control system, but without the cash penalty provision. On no occasion has it proved necessary to actually remove concrete from any of these structures.

Generally, concrete suppliers have been responsive to requests to adjust cement contents based on early age analysis. However, there have been frequent occasions when the strength provided, assessed as above, has fallen below that strictly required, for extended periods, by 1 MPa (145 psi) or less.

Such minor deficiencies have no structural significance but do waste time in repeated requests and reports and arguments with concrete suppliers (who are ever optimistic that the 7- to 28-day strength gain will improve on current production). Suppliers complain that precise enforcement is unrealistic, yet without strict controls, deficiencies would no doubt tend to gradually increase. A cash penalty as proposed would avoid all need for such argument. The deficiencies would be acceptable with the penalty paid, but it is suspected that deficiencies would rapidly disappear in such circumstances.

There have been suggestions that, in fairness, penalty clauses should be balanced by bonus clauses. This is not recommended because excess strength beyond that specified is of little benefit to the owner and may be detrimental. The type of cash penalty clause advocated here is a real benefit to the good concrete supplier. He can aim at the mean strength truly needed without restriction. If he slightly miscalculates, the penalty is very moderate and involves no cost of delays or further investigation. He is defended from unfair competition by less competent or less scrupulous competitors. Finally, he can include his own bonus in his pricing if he wishes.

Conclusions

It is concluded that a cash penalty of twice the cost of the cement or admixture deficiency can be accurately established by the analysis of a group of 30 consecutive test results. Such a penalty would be capable of regulating concrete strength with fairness. The system would result not only in an improved degree of contractual compliance but also in a cooperative attitude in day-to-day control between the contractor and the supervising engineer. It would provide an effective incentive to improve control, which would, over a period, produce significant improvements in concrete production techniques.

11.2 WHAT IS ECONOMICAL CONCRETE?

This section appeared in *Concrete International* (Day, 1982a). It is quoted verbatim as Day's views have not changed. Permission granted by the American Concrete Institute to reproduce it here is gratefully acknowledged.

The question "What is economical concrete?" may seem a ridiculous question, but consider the example of the Rialto project in Melbourne. This project is very unusual in that the concrete supplier, the builder, and the eventual owner were one and the same. It involved 6000 m³ of a 60 MPa (8700 psi) grade, which was the highest grade of concrete specified for such a project in Australia at that time. This was only 6% total concrete quantity. Accordingly construction started with a very conservative mix that actually provided a mean strength of over 80 MPa (11600 psi) and a characteristic strength of approximately 75 MPa (10875 psi). Considerable cement content reductions (say, 100 kg/m³ [170 lb/yd³]) were clearly possible but no reduction was in fact made on the following grounds:

1. The possible saving of $60,000 was trivial compared to the total project cost of several hundred million dollars.
2. The huge strength margin virtually ensured that there would be no delays due to strength problems.
3. The very high early-age strength permitted early stripping with no concern for damage, weather conditions, need for intensive *in situ* or early-age testing, and so on.
4. The additional safety margin against any unexpected factors was also of some value.

As another example, Australia's billion-dollar Parliament House is a major concrete structure, containing about one quarter million cubic metres of concrete. At around $25 million, the cost of the concrete supply represented about 2.5% of the total cost. It really would not matter very much if this cost increased by 5% to 2.63% of total cost.

Of course, the extra cost in the case of the Rialto would be a little less trivial if the same argument were applied to the whole of the concrete in the project, but the real point is that this attitude could never be taken by an independent concrete supplier because the cost would probably exceed the entire profit margin. The strength margin (but more likely 5 MPa [700 psi] than 15 MPa [2000 psi]) could therefore only come about by either the owner specifying a higher grade or the builder ordering a higher grade than specified. Either party might take this action on the basis of expediting construction or at least of avoiding any risk of delay. In fact the best way of organising this is for the owner to specify a higher strength but to impose a cash penalty rather than rejection or further investigation for strength shortfalls of up to 5 MPa (700 psi) (or whatever margin has been allowed). The same effect could be obtained by offering a bonus for excess strength (of course within a strict limit) and not raising the specified strength.

The benefits accruing from the proposed technique (of specifying a higher strength than strictly necessary and providing a cash penalty for strength deficiencies within the margin) would be as follows:

1. A relaxed attitude to minor strength deficiencies by the owner
2. A keener attitude to minor strength deficiencies by the concrete supplier
3. A smoother running project
4. The provision of better concrete, probably at only a very marginal overall cost increase

There is yet one remaining possible turn of the screw of increased strength margin. This is to obtain the extra margin not by specifying a higher strength but by specifying a lower percentage defective at the original strength. This would have the effect of putting a higher premium on low variability and could be a substantial factor in discriminating in favor of better producers and so providing a beneficial pressure toward improved performance by the industry. If a strength increase of the order of 5 MPa (700 psi) is desired, it would amount to around 1.5 times the standard deviation. In most of the world, a 5% defective level is used, so that a mean strength of specified strength plus 1.645 times standard deviation is required. Raising the margin to 3 times standard deviation would go close to the 1 in 1000 defective level (mean 3.09 standard deviation) and would mean, for a typical 3 MPa (435 psi) standard deviation, providing a margin of 9 MPa (1300 psi) between mean and specified strengths. The margin would vary between 6 MPa (870 psi) and 12 MPa (1740 psi) from the best concrete producers (SD of 2 MPa [290 psi]) to the worst we should tolerate (SD of 4 MPa [580 psi]). With such a pressure to improve, it is likely that in 5 or 10 years time, we would find the good operators down to below 1.5 MPa (220 psi) SD (margin of around 4.5 MPa (650 psi) as currently typical) and the rough operators out of business. Perhaps an intermediate solution would suffice. A margin of much less than 5 MPa (say, 2 MPa [300 psi]) is probably quite adequate for the operation of a cash penalty system and this would be provided with an SD multiplier of 2 (giving around the 2.5% defective level of 1.96 SD). Incidentally it is time we stopped thinking of SD multipliers primarily in terms of permissible percentage defective. The real grounds on which they should be selected is the relative value we place on mean strength and standard deviation in assessing concrete quality (on this ground, a multiplier of 3 is highly desirable). The relationship between the desirable mean strength (or the 10%, 5%, or 0.1% defective level) and the strength used in structural design calculations should be a subsequent rather than an initial decision but is clearly an independent decision.

Interestingly, the cost of the additional strength margin now being proposed (or more) has often been incurred in the past by the specification of z 20 MPa (3000 psi) characteristic strength together with a minimum cement content requirement of the order of 300 kg/m³ (500 lb/yd³). There is however a very substantial difference in the results of the two specification bases. Whilst the former offers distinctly better concrete, a smoother running project (due to the cash penalty basis) and a pressure toward a better performing concrete industry, the latter offers scope for cheating on cement content, for the use of substandard aggregates and oversanded, high shrinkage mixes and, most important of all, a removal of any incentive for the technical competence of producers.

There are two important provisos that should be made in advocating cash penalties and greater emphasis on standard deviation:

1. The standard deviation (and the mean strength, but that is much easier) must be accurately determined.
2. The cash penalties (which may be described as "liquidated damages" or "provision for reduced durability" or formatted as a bonus clause rather than a penalty) should be very moderate, only about twice the cost of the additional cement that would have avoided any penalty (i.e., about 10 kg/m³ per MPa [12 lb/yd³ per 100 psi] of deficiency so, in Australia, about a $2 penalty per MPa of deficiency).

The requirement for an accurate SD is easily satisfied under a cash penalty system because it is not necessary to identify which concrete is slightly understrength, only how much and how defective. Therefore the penalty can be levied on the concrete represented by 30 consecutive results with great accuracy.

Does anyone have a convincing counter argument? If not, how long do you think it would take to implement this proposal? 5, 10, 20 years? It may be of interest that the outline of this argument was advanced in papers published by Day in 1959 and 1961.

11.3 HOW SOON IS SOON ENOUGH?

The first edition contained a 21-page account of an investigation using a massive computer analysis of synthetically generated data to clearly establish the superiority of cusum analysis over any other system known for the early detection. That section of the first edition is not repeated here but is available on the website.

The two most significant points arising were as follows:

- No computer analysis is as efficient and reliable as the eye examination of a cusum graph in detecting a small change in mean strength.
- The mathematical significance of a downturn in a single variable (i.e., strength) is in any case immaterial when the significance of the downturn is correlated to simultaneous changes in other variables such as slump, temperature, and density.

The economic value of a more efficient analysis system is briefly compared to that of other factors affecting the attainment of the desired concrete quality, such as better equipment, more skillful personnel and higher testing frequency. It is pointed out that a more efficient detection system is equivalent to a higher testing frequency in achieving early detection.

It is shown that the average number of results required to achieve detection of a change is directly proportional to the standard deviation of those results. Since early detection in turn enables a reduction in variability, a self-intensifying cycle of variability reduction is commenced.

The questions of early-age and accelerated testing; of monitoring batching performance; of analysing related variables such as slump, density, and temperature have been addressed elsewhere in this volume. For the purpose of this investigation it was assumed that a continuous string of test results is being received and converted into predicted 28-day results. The relative efficiency of the different techniques in detecting a downturn in such a string of results was examined.

Based on 40 years of plotting quality control charts for concrete, Day's assumption is that the downturn is usually a sudden event or step change rather than a gradually worsening trend. To highlight this point, a simulation was conducted to automatically produce a string of 100 random but normally distributed results of any selected mean and standard deviation. It then added a further 30 results of the same standard deviation but a lower mean. This enabled examination of the performance of a control system in respect of whether it raised false alarms during the initial stable period of 100 results and how long it took to detect the imposed change point at the 100-result mark. The results were automatically analysed by up to six different detection systems at a time and the results reported as follows:

1. The number of results prior to a false alarm in the first 100 results, if the number is 100, there were no false alarms.
2. The number prior to the first detection of change after the imposed change point, if the number is 30, there was no detection.

The best detection system is not necessarily the one that shows the lowest average number of results to give a detection. Any type of system can be made more sensitive by narrowing its limits, at the cost of experiencing more false alarms. It was not considered sufficient to find that one system was extremely good at detecting changes but gave many false alarms, while another gave few false alarms but was a poor detector. It is certainly of interest to compare the relative severity of different national codes but the authors' primary interest is in finding the most efficient way of detecting a change. The exercise was therefore repeated after adjusting the nominal specified strength so that each system gave similar false detection frequencies when assessing the same concrete.

It was found to be important whether the adjustment was in the form of a constant or that of a multiplier of the SD. The various national systems often incorporate a fixed adjustment, for example, ACI 214 requires not more than 1 in 100 results to be more than 500 psi (3.45 MPa) below the specified strength and BS 5328 requires the running mean of four results to exceed the specified strength by at least $3N/mm^2$ (3 MPa). This investigation has shown that unless such adjustments are expressed in terms of a multiple of the standard deviation, the systems would give a substantially different relative performance according to whether the production was at high or low variability. Another aspect of system efficiency is the use of multiple criteria. A system can be made to give a better ratio of correct detections to false alarms by composing it of several subsystems running in parallel. In this case the better performance is obtained at the cost of a more complicated criterion, a larger program, and slower operation. With computer assessment, these costs would be negligible compared to increasing physical testing frequency and it should be realised that a more efficient analysis system has as much value as additional testing. For example it would be possible to analyse results using a combination of all the systems and to accept that a downturn had occurred when one was detected by any two, or any three, of the nine systems shown. This would no doubt give both a better detection rate and less frequent false alarms. However, the improvement would probably be relatively small since false alarms are frequently due to aberrations in the results affecting several systems rather than to aberrations in one of the detection systems. (In this respect it would be of value to people involved in concrete QC to examine a selection of the data generated for this investigation in order for them to realise the extent to which apparently convincing downturns in a set of results occur as a result of normal statistical variation.)

The real reason militating against the multiple criteria approach is that they must still be suitable for the average user. Complication must be avoided as far as possible, both to ensure comprehension by all concerned in their enforcement and to avoid the much greater effort of examining compliance by manual calculation by persons not having computer knowledge or facilities.

11.3.1 Relative performance of the systems

All the systems, except ACI 214, are nominally directed toward assuring a characteristic strength that 95% of results will exceed. Therefore, that characteristic strength is given by the mean minus 1.645 times standard deviation, that is for this exercise, 35 – 1.645 SD.

In the case of ACI 214 the requirement is for only 90% of results to exceed the specified strength. Therefore that strength in this exercise becomes 35 – 1.28 SD. However in the adjusted limit section, the ACI system is still comparable as what is reported is the amount of adjustment required.

It can be seen that both the ACI and the UK systems, in their original forms, give rapid detection of a downturn but also give a high rate of false alarms. The Australian system on the other hand appears unduly lenient. The numerical cusum was adjusted to comply with the 70/80 false

Table 11.1 National criteria as in national codes

	ACI214	AS3600	BS 5328	N Cusum
False alarm frequency	52.36	93.6	46.81	70.74
Average detection delay	1.75	12.90	2.64	4.11
Maximum average detection delay	8.06	20.15	7.26	10.54
Adjusted (by Constant Margin) to comparable false alarm frequency				
Adjustment in char strength (MPa)	1.75	−0.60	6.50	NA
False alarm frequency	63.8	64.5	77.8	71.2
Maximum average detection delay	17.6	20.3	22.5	16.0

alarm frequency during the process of selecting the deduction margin and detection limit. This was done as a separate exercise using the techniques of this investigation in which a large range of margins and limits were tried (in sets of six) to find the most efficient combination.

The basic techniques embodied in the national codes (individual result limit and limits for running means of 3, 4, 5, and 30) were also separately examined (Table 11.1). This was necessary because some of the combinations were optional and also to avoid concluding that the code incorporating the largest number of individual criteria (ACI 214) was necessarily the best.

11.3.2 Visual cusum

In the initial stages of the investigation, hundreds of graphs of the run of 130 results were examined. It was noted that the basic cusum graph almost invariably showed a quite distinct downturn at the exact point of the artificial downturn, even when the drop ratio was so small that the numerical system detection efficiency was poor.

It should be noted however that this is far from the same thing as concluding that the detection efficiency of the basic cusum is almost perfect. The technique looks better in retrospect than it does in genuine use. Examination of the overall 130 result trial tells nothing of how many of the small false downturns in the cusum graph might have been mistaken for the real downturn or for how long an operator might have regarded the real downturn as such a false one. So, while the keen and experienced operator using cusum graphing will already have acted before the detection system provides a signal, the less experienced operator will be glad of the confirmation provided by the system and the less keen operator will be prodded into action.

What is clear is that, on looking back after concluding that a downturn has occurred, the basic cusum graph will show exactly when that downturn occurred. This is very valuable information because the same logic applies to any other variable for which a cusum graph is drawn, and therefore it is usually easy to match up cause and effect.

11.3.3 Numerical cusum

The previous mean value is subtracted from each result and if the difference numerically exceeds a selected margin, the difference (less the margin) is accumulated in a register. If the accumulated total exceeds a selected limit, a detection has occurred. In practice positive and negative registers are maintained (because detection of an upturn means that cement can be saved, which is a further reason to prefer numerical cusum) but for the current exercise, only a negative register was maintained.

For any selected margin, a limit can be chosen to give whatever frequency of false alarms is considered acceptable. It is conventional to choose a margin of about half the minimum change it is desired to detect. If this is considered to be $0.5 \times SD$, then $SD/4$ might be the chosen margin. The investigation reported started with a margin of $SD/3$ and a limit of $4 \times SD$, but after comparative trials the best results were obtained with a margin of $SD/6$ and a limit of $5.5 \times SD$.

The use of a numerical cusum in this way is exactly equivalent to using a graphical V-mask technique (Devore) as is used in the United Kingdom.

11.3.4 Assessment of alternatives

On average and after adjustment to a comparable false alarm frequency, the running mean of five gives the quickest detection. However the numerical cusum follows close behind and is better at detecting very small drops. Numerical cusum is also more directly aimed at detecting change from a previous situation rather than infringement of a specified limit. Since a producer would be ill-advised to work right down to the limit, the latter is likely to be the more useful feature. Numerical cusum is also equally at home in detecting upturns and this is important to the producer. Of course a running mean of five can be adapted to all these purposes, but this is not often done.

The national systems are not strictly comparable as they have different intended methods of application. The American ACI 214 publication sets out a range of possibilities together with several pages of excellent advice and information with the objective of allowing specifiers to make their own informed decisions. It also includes a recommendation to maintain control charts and detailed advice on how to do so.

The British BS 5328 condenses its unequivocal requirements into a small table and four carefully chosen sentences. To be fully comparable with the ACI system it would also be necessary to make reference to the requirements of the British Quality Scheme for Ready Mixed Concrete, which is an industry based self-regulatory body and recommends cusum control charts or an alternative counting rule system involving not more than eight consecutive results below the previous mean.

The Australian system provides a rule by which concrete producers are required (regardless of individual project specifications) to regulate the whole of their production. It then also provides a rule by which individual projects can check the quality of concrete received by that project.

In comparing the requirements it should be remembered that the British code is anticipating approximately double (4 to 6 MPa) the standard deviation normal in Australian capital cities (2 to 3 MPa), with the United States covering a larger and intermediate range. It could also be said that the Australian code is designed to avoid unfair condemnation of the producer and allow full benefit for the attainment of low variability, while the British code is aimed solely at providing near certainty that the supply of substandard concrete will be eliminated in all circumstances. It appears that the carrot may be currently showing greater benefit than the stick.

The use of a minimum required strength for any individual specimen has good and bad points. It is reasonable to put a limit to the downward spread of results, which could be obtained with very high variability concrete while still providing a mathematically acceptable mean. However, test results are subject to error and an individual specimen criterion can require action on the basis of a badly made test if not intelligently administered, and the authors' experience is that such matters are often not intelligently administered.

The use of a fixed lower limit for individuals may have its merits but the use of a fixed numerical limit for the running mean of a set of four, as in the UK Code BS 5328, has the unfortunate effect of severely limiting the financial benefit obtainable from good control. As previously noted, any kind of requirement involving a constant produces distortions in performance over a range of SD values.

One final answer to the how soon question must be "before anyone risks their neck". It is quite possible to assess concrete quality within 24 hours and it is probably legally, and certainly morally, indefensible not to do so prior to prestressing, early demolding, jump form movement, and so on.

11.3.5 Other significant considerations

Where cost competition is negligible, it is easy to provide a large safety margin totally avoiding failures. In these circumstances a highly tuned control system may not be essential but is obviously affordable.

Where cost competition is severe, a control system that can detect a shift in mean strength of as little as 1 MPa (150 psi) within 2 or 3 days of its occurrence may be an excellent investment. Where operating conditions and materials are very stable, the additional cost of early age testing may not be justified. Seven-day testing has the advantage that on detection of a suspected downturn a reservoir of test specimens from 1 to 6 days age is available and can be immediately brought forward for test to confirm or negate the change. This is providing one is sufficiently knowledgeable (and has done the necessary prior investigations) to correctly interpret results at a range of early ages.

The control process should be considered as a whole, ensuring value for money in several different types of expenditures, for example:

1. Batching equipment
2. Quality of testing
3. Frequency of testing
4. Computer equipment
5. Computer software

The ability to work to a 1 MPa (150 psi) lower mean strength for a given specified strength is worth about 5 kg of cement per cubic metre (8.4 lb/cu yd). This is a sufficient saving (on high volume production) to pay for a very elaborate control system. The ability to detect a downturn in strength a day earlier may avoid a major penalty. It may also justify a lower safety margin.

It should be noted that all criteria relate to the standard deviation of results. Lower variability concrete is easier to control more precisely. As already noted, this is not tautology but a recognition of a multiplier effect of control improvement. A reduction of 1 MPa in standard deviation makes a direct difference of 1.28 or 1.65 MPa to the required target strength (depending on whether the specification is based on 90% or 95% above). It will make at least a further 1 MPa reduction in the strength margin required for the detection of a change. Improved quality control may also be a major sales point. The standard deviation of the concrete strength is obviously affected by the quality and effectiveness of both the batching system and the testing process, as well as by the variability of input materials.

The frequency of testing is an important cost factor to be weighed against the quality of testing; the securing of additional data, such as slump, concrete temperature, and density; and the cost of result analysis. The cost of elaborate analysis is rapidly reducing compared to that of physical testing and an increase in one can justify a reduction in the other.

The ability of a control system to combine results from many different grades of concrete into a single analysis can be equivalent to a several-fold increase in testing frequency.

The time between a downturn and its detection and rectification is also affected by the age at test. The days in which mix revisions were based on 28-day test results are hopefully gone, but the choice of test age in the interval of 1 to 7 days is open to consideration. In temperature-stable tropical conditions, 3 days is a good choice. Depending on the protection provided to the specimens, and on the time of collection, a 3-day strength may be too variable in other climates. Further options are to use accelerated specimens or to measure thermal maturity in order to obtain a result at 1 to 2 days.

A consideration of the above factors makes it clear that

1. Except in very low volume situations, there is ample saving in cement cost to offset a high standard of control.
2. The cost of computer analysis with a good class of computer and software is modest compared to other factors in achieving timely control of concrete quality.

Chapter 12

Troubleshooting

There are several aspects to troubleshooting in concrete technology. One of them, separation of its costing from that of quality control (QC), was raised in the first edition and is repeated here:

Another is the examination of existing structures with a view to repair. However we note the following points to consider:

- The field is a very extensive and rapidly developing one and, to provide good professional service, requires that the practitioner keep fully up to date with a myriad of constantly changing techniques and proprietary materials. Repairs to concrete structures are very often temporary (usually unintentionally) and may provide only a short-term cosmetic effect at considerable expense.
- Clients are often unwilling to consider the expensive solutions that may be necessary to achieve a degree of permanence.
- Even the specialists have difficulty in establishing which of several competing repair proposals represent best value for money (or whether any proposal offers good value).

Younger readers should note that this field is likely to absorb something like half the total expenditure on concrete structures in the next few decades. It is also likely to generate distinctly more than half the profits to be made out of concrete technology in this period. This is because many asset owners are more willing to pay for cure than for prevention (even if timely prevention costs a small fraction of the cure).

The authors have from time to time helped to sort out problems with concrete still in the production stage. Advice on the procedure to follow seems desirable since the kind of action necessary in many (but not all) such situations is reasonably easy to learn (compared to repair), and since even relatively amateur attempts to follow the advice given are likely to be beneficial, even if not necessarily optimum.

The first action must be to establish exactly what the problem is. Some possible problems are

- Inadequate strength
- Difficulty in pumping/blockage
- Inability to compact
- Unsatisfactory appearance
- Excessive segregation or bleeding
- Inadequate retention of workability
- Failure to set or stiffen sufficiently rapidly
- Presetting cracks or later age cracks
- Excessive cost of imported materials
- Excessive variability

Possible problem sources are

- Unsatisfactory aggregates
- Unsuitable mix design
- Poor testing (including sampling, casting, and curing of specimens)
- Cement or pozzolan quality
- Unsuitable admixtures or admixture usage

Data to request (having relevant past data available on arrival can often shorten the investigation by a day or more):

- Mix details
- Aggregate gradings
- Delivery dockets
- Concrete test records (including times, temperatures, and specimen collection details)
- Cement test certificates if available
- Cores and failed test specimens to inspect

Of course it is desirable that records go back to a period before occurrence of the problem if possible. Where aggregate testing records seem inadequate, a rapid visit to the stockpiles is desirable before (further) change occurs. Segregation of coarse aggregates, silt content of the sand, and contamination with subgrade material by front-end loader are items to look for.

12.1 STRENGTH, PUMPABILITY, AND APPEARANCE

12.1.1 Inadequate strength

The typical steps taken when called in to investigate problems may be of interest. The steps are

1. Restore strength to a safe level so work can continue while investigating. Cement content adjustments should always overshoot when increasing and undershoot when reducing. Use 8 to 10 kg per MPa to adjust upward or equivalent water reduction, 4 kg per MPa to adjust downward. If adjustment gives cement content over 500 kg use 500 kg plus 2 kg of fly ash (for each 1 kg of cement not added), or 0.5 kg silica fume, or 100 ml superplasticiser.
2. Start casting at least four, perhaps six, test specimens per sample. Test at 2, 3, 7, 28 and perhaps 56 days. Assume gain in megapascals will remain the same with the revised mix. In default of prior data, conservatively assume that strength will increase 33% from 2 to 3 days, another 33% from 3 to 7 days and 10 MPa from 7 to 28 days. Substitute actual figures as soon as available. Testing one specimen at a particular test age can be a problem if testing quality is an issue.
3. Draw cusum graphs of strength (at all available ages), density, concrete temperature, slump, 7- to 28-day gain (for example). If data is available, cusum graphs of sand silt content or specific surface should also be drawn on the same presentation. A cusum of average pair difference between pairs of specimens from the same sample will show whether there has been a deterioration in quality of testing (an average pair difference in excess of 1.5 MPa is an indication of poor testing quality). Such graphs will usually show when the problem started and what caused it.
4. Examine batching records (assuming a computer-operated plant that records actual batch quantities) before and after the downturn for signs of cement shortfall or aggregate, especially sand, overbatching.
5. Calculate MSF (mix suitability factor) using formulas in Chapter 8. MSF is a measure of the sandiness of the mix taking into account sand grading, sand percent, cementitious material content, and entrained air. Calculate water content using formulas in Chapter 8. Is actual water content really known? An MSF in excess of 30 represents oversanding and high water requirement unless for flowing, superplasticised concrete.
6. Calculate strength according to one or more of formulas in Chapter 8. If this agrees with strength obtained/being investigated, then high water content is the explanation and the reason and cure are obvious (may be any combination of high MSF, silt in sand, concrete temperature, high slump).
7. If calculated water or strength does not agree with actual, recheck sand silt percent and grading. Check concrete density, as this will confirm water, air content, or compaction of test specimens. The water content is the major separating factor between alternative directions of investigation. If water is the end cause, then the basic cause is likely to be in the area of dirty or finer sand, high sand content, high slump,

or high concrete temperature. If water is not the cause, then the basic cause is likely to be in the area of poor testing (including sampling, compaction, curing, capping [if cylinders], defective or badly cleaned/ assembled molds [if cubes], centering, load rate, etc.), or of cement quality or quantity.

8. Use the AASHTO T318 Standard method of test for water content of freshly mixed concrete using microwave oven drying to help confirm the actual total water content from all sources.

12.1.2 Poor workability/pumpability

Generally the causes are an excess or deficiency of fine material, a gap in the grading, or an excess or deficiency of fluidity.

1. Does the concrete bleed? If so, there is either a gap in the grading, a deficiency of fine material, or excessive fluidity. If the concrete pumps reasonably at the start, but will not restart after a delay, this is often due to bleeding.

2. Using Ken Day's MSF, the value of this must be at least 24 to 25 for pumping to be possible. The higher the desired fluidity, the higher the MSF value will have to be; however, values in excess of 32 will exhibit excessive friction unless superplasticised to high slump.

3. Draw a graph or produce a table of individual percentage retained on each standard sieve. Ideally all sieves below the largest will have similar percentages of around 7% to 10%. One size missing may not be fatal if those on either side are normal. Any two consecutive sieves with a combined total retained of less than 7% would be a potential problem. More than 20% on a single sieve finer than 4.76 mm might also create a problem in pumping.

4. Is there at least 300 kg/m³ of material passing the 0.15 mm sieve (including cement)? If not, additional fines may be needed as fine sand, crusher fines, fly ash, or cement.

5. If the (single) sand is so coarse that more than 55% (perhaps 50%) of it is necessary to provide an MSF of 25 there is likely to be a problem with bleeding, segregation, and pumpability. Additional fines as in item 4 are necessary.

6. Air entrainment, fly ash, and silica fume (in increasing order of effectiveness) are effective suppressors of bleeding and so assist pumpability. The authors have witnessed huge foundations up to 4.5 metres deep filled with flowing concrete and containing 40 kg/m³ of silica fume, which exhibited no bleeding whatever.

7. Although nothing to do with mix design, it should be borne in mind that it is pressure that causes a problem in pumping and faster pumping requires higher pressure. Also a delay caused by a gap in deliveries

is an aggravating factor. Therefore, if pumping problems are being experienced, pumping more slowly and ensuring that one truck is not emptied before a replacement arrives may assist.

8. The use of a priming slurry that is too wet is a common cause of pump blockage. This is because the focus is on wetting the pipeline not segregating the concrete that follows. The optimum slurry mix should be cohesive with the similar composition as the concrete but without the aggregates.

12.1.3 Unsatisfactory appearance

Unsatisfactory appearance may be due to inept placing, poor formwork, or many other factors that are beyond the scope of this book. However, it is also often due to bleeding, the remedies for which were covered earlier. If bleeding happens at all, it often results in a flow of water up the face of formwork, leaving clearly visible signs. A slight formwork leak (just of water) can cause an internal surface flow of water over an area of more than a square metre and result in a large black stain, known as a hydration stain.

Presetting or plastic cracks—There are two kinds of presetting cracks with diametrically opposite causes: settlement cracks and evaporation cracks.

Plastic settlement cracks—Plastic settlement cracks result from settlement of the concrete due to loss of bleedwater. In settling, the concrete "breaks its back" over anything resisting settlement in one location and not another, for example, reinforcing bars, cast-in plumbing, or sharp changes in depth of section. Measures to avoid bleeding were dealt with earlier.

Plastic shrinkage or evaporation cracks—Plastic shrinkage or evaporation cracks result from evaporation of water from the surface layer of concrete beyond the rate of bleed from the concrete to replace the loss. If a concrete has very low bleeding (e.g., silica fume concrete), it is susceptible to such cracks and measures must be taken to avoid evaporation (e.g., use of an aliphatic alcohol evaporation retardant such as Confilm, a sheet material such as polythene, or a mist spray of water drifting across the surface).

Thermal stresses—Another frequent cause of early-age cracking is thermal stress. This can be reduced by substituting pozzolanic material for cement in the mix design and reducing placement temperature. However action other than mix change may be needed, such as avoiding restraint to thermal shortening (in the case of long slabs); reducing temperature differentials between the element and the substrate or within the element. It is important to know the likely cause of cracking: internal or external restraint.

Autogenous shrinkage—Autogenous shrinkage should not be forgotten as a cause of early cracking in low water to cementitious materials (w/cm) ratio mixes. Hydration removes free water in the concrete causing tensile forces to develop within the capillary network. Unlike drying shrinkage, it can occur in spite of any measures taken to reduce or prevent evaporation. It is also a through section shrinkage that can exacerbate thermal shrinkage problems. Autogenous shrinkage can be a particular concern with low w/cm ground-granulated blast-furnace slag (GGBS) concrete.

As autogenous shrinkage is due to tensile stresses developing in the capillaries, which requires the formation of a meniscus. Therefore ponding with water during early curing period is a good way to limit autogenous shrinkage, at least in the surface layer.

12.1.4 Excessive variability

The first thing is to establish whether the variability is in the concrete or in the testing. Two places to look are the average pair difference in the 28-day results and the range of densities of test specimens from the same sample of concrete. The average pair difference should desirably be below 1.0 MPa and densities should not have an average range exceeding 50 kg/m³. However, calculated densities may vary through inaccurate measurement of specimens rather than variable compaction or segregation, and this would have no effect on strength variability.

A second place to look is at multivariable cusum graphs of strength and other variables. If slope change points in strength correlate with those of other variables, the cause will be clear. Direct plots of multiple variables will show whether individual high or low results have an explanation. If there is no explanation, and especially if 7- and 28-day results do not correlate, testing would be suspect.

Having established that the variability is actually in the concrete and not just the testing, batch quantity records should be available if batching is by a computer-operated plant. It should not be overlooked that the correct quantities may be weighed out but may be insufficiently mixed to give uniformity. There have also been examples of short central mixing times (prior to further mixing by agitator trucks) that have not permitted time for all the metered admixture to enter the mixer. Similarly part of a particularly critical ingredient such as silica fume may hang up in the batching skip from time to time and finish up in the next load. Alternatively silica fume may not be properly dispersed greatly reducing its effectiveness.

12.2 CAUSES OF CRACKING IN CONCRETE SLABS

The causes of cracking in concrete are sufficiently well known to permit their automatic diagnosis in most cases. Day wrote an expert computer system for this purpose. An expert system is a computer program that asks questions of a user in order to be able to diagnose the cause of the user's problem; the better ones are also able to explain why the particular question is being asked, on request by the user.

The first question to be asked is the age of the concrete at cracking. If the age was less than 10 hours, the crack would be classified as a plastic or presetting crack caused by either excessive evaporation from the surface or by restrained or differential bleeding settlement. If the age was more than 10 hours but less than 48 hours (and especially if the crack occurred in the early morning following pouring) the crack would probably be a thermal contraction crack. If the age exceeded 2 days (and was after termination of covered or moist curing if any) it may be due to drying shrinkage.

To determine whether plastic or presetting cracks are caused by evaporation or settlement, questions are asked about the shape, size, and orientation of the crack, and about whether the concrete bled substantially or was subjected to drying winds and low humidity. Plastic shrinkage cracks may be quite wide on occasions, but they are usually short and randomly orientated. However, they can sometimes be concentrated in an area of the slab that is more exposed to wind and can form parallel lines. In the latter case they may be more difficult to distinguish from settlement cracks occurring over a steel mesh, except that it would not be likely that evaporation cracks would be parallel to the direction of the mesh, or at the same spacing. As already noted the settlement cracks can occur over reinforcing bars, installed plumbing, or the like. They can also occur at lines where the section deepens, such as dropped capitals for columns, haunched beams, or the edge of thickened areas of a slab (waffle slabs, slab/beam).

A classic situation for thermal cracking exists when a concrete wall is poured between restraints. The restraints may be a heavy foundation beam with starter bars or substantial columns with projecting reinforcement. When a wall in such a situation is poured on a warm afternoon using a mix rich in a Portland cement the width of the crack to be anticipated on stripping next morning can be calculated if a maximum reading thermometer is inserted. Such cracks are often widest at the base, near to the restraining foundation beam, and taper away to nothing 2 or 3 metres up the wall.

A commonly encountered situation is where a crack runs parallel to, and often close to, a sawn control joint. It is easy to see that either the joint was not deep enough to be effective or, more likely, it was actually cut after the slab had already cracked, although perhaps before it had opened sufficiently to be noticeable.

Another useful distinguishing test is to place a straightedge at right angles across a crack. If the straightedge will rock, this indicates that the slab has deflected and therefore that the crack was probably caused by subgrade or formwork movement, or structural inadequacy in the case of suspended slabs.

Where cracks are three pointed, they are usually caused by a swelling or settlement resisting rock immediately below the junction of the cracks, for example, a "floater" in a soft subgrade subject to moisture movement.

In the case of suspected thermal cracks, it is useful also to check whether the concrete had a high cement content, making it likely to generate more heat, whether it was poured on a hot afternoon followed by a cold morning, and whether there was a delay in pouring, which could have allowed the concrete to heat up while kept waiting in the truck.

Surface crazing occurs when the surface layer shrinks relative to the body of concrete below it. This can be caused by allowing the surface to dry or cool quickly and is more likely when a high shrinkage surface layer, rich in cement paste and fine sand and of high w/c ratio, is present.

There is an almost universal tendency to use quality control personnel for trouble shooting of the above nature. This may be a reasonable use of any spare time, but it is important to ensure first that it does not disrupt the QC routine and second that such work is separately costed from QC. This is because the economic justification of QC should be clearly established as it otherwise tends to be regarded as a luxury item, first in line for cutting in hard times. Troubleshooting in general is not QC, indeed it may be the result of inadequate QC, and it is rarely cost saving or revenue generating. Many QC departments (not only in the concrete industry) have been axed or decimated through a failure to recognise this.

Chapter 13

Concrete future

13.1 REMOVING IMPEDIMENTS TO THE MORE SUSTAINABLE USE OF CONCRETE

James Aldred[1]

13.1.1 Introduction

The concrete industry is keen to position itself as an integral part of sustainable construction. Indeed, it is hard to think of sustainable development for the growing global population without thinking of concrete as the primary building material for structures and infrastructure. However, there are many impediments to the more sustainable use of concrete within projects. In fact, the contagion of excessive risk aversion and regulation sweeping the industry appears to be on a collision course with sustainably meeting the needs of the present. Sometimes even so-called sustainable requirements cobbled onto existing specifications may result in reduced sustainability.

13.1.2 Environmental policy

In his speech at Harvard University in 1947 George C. Marshall said: "An essential part of any successful action ... is an understanding on the part of the people ... of the character of the problem and the remedies to be applied. Political passion and prejudice should have no part". Arrhenius suggested that fossil fuel combustion might eventually result in enhanced global warming as far back as 1896. Yet it was thought that human influences were insignificant compared to natural forces and that the oceans acted as such vast CO_2 sinks that there would be no net accumulation in the atmosphere. The decrease in global annual temperature from the 1940s to the 1970s in spite of significant increases in carbon dioxide shifted the

---

[1] Adapted from the opening paper given by James Aldred at the Concrete in the Low Carbon Era Conference, Dundee, UK, 2012.

focus to global cooling. The increase in global temperatures in the 1980s resulted in modeling by Hansen (1988) and others that suggested alarming temperature rise if the output of anthropogenic carbon dioxide were to continue unchecked. These predictions on expected temperature rise provided an objective basis for assessment of the anthropogenic global warming (AGW) hypothesis. The problem arose with the rush to political action before confirmation of the hypothesis. Bypassing the normal scientific debate and assessment process prevented the development of a true understanding of the problem and exploration of possible remedies. As a result, the AGW discussions have been full of "political passion and prejudice" from the outset.

The AGW/CO_2 issue has now become a divisive impediment to the crucial sustainability issues of resource depletion and damage to the natural environment. At a recent meeting on adapting to the carbon tax in Australia, which came into effect this year, the use of recycled aggregate was considered not a viable option as it had the same carbon footprint as virgin aggregate. However, increasing the efficiency of our built environment to minimise energy consumption, reducing the requirement for virgin resources, facilitating renewable energy and developing truly sustainable communities are examples where the concrete industry is making considerable progress. If we were to achieve realistic targets in each of these areas, we would profoundly reduce fossil fuel consumption and CO_2 production without the need for carbon taxes and alike.

Living on a planet with 7 billion people and limited resources, there is virtually universal agreement on the importance of resource depletion and damage to the natural environment, regardless of one's opinion on AGW. Refocusing attention back onto the primary sustainability goal of meeting the needs of the present without compromising the ability of future generations to meet their own needs would appear the best way to harness our collective efforts for optimum benefit.

13.1.3 Risk aversion

Problems with risk aversion arise from limited data and engineers erring on the side of caution. Adjacent to a major project, there was a plan to extend a local waterway. It was suggested that this might raise the water table, although it was deemed by one of the consultants involved to be a "50:50 call". Even though the construction was underway, the consultants decided to increase the height of the pile caps over the massive site at significant time, cost, and materials. If the announcement to extend the waterway had been delayed by a few months, the project would have been constructed based on the original design. The project would have had to deal with the effect of the waterway on the water table in the same way as the neighboring structures, which had been already completed and could not have made this 50:50 call.

Another major project had a number of engineering firms designing the different phases. The proposed reinforcement for the basement rafts varied by more than 300% despite each designer being given the same specification guidelines. One designer proposed massive perimeter restraint to limit predicted lateral movement and the others did not. How could there be such a range of different technical requirements to build essentially the same structural elements? One factor was the issue of calculating the limiting crack width. The highest reinforcement volume proposed was required to reduce the limiting crack width below the 0.2 mm nominated in the design brief on the basis that it would need to be a watertight structure when the waterproofing membrane eventually failed. Another important difference among the proposed designs was the expected stress caused by long-term shrinkage. The American Concrete Institute (ACI) guidance largely ignores shrinkage after one year on the basis that the rate of shrinkage will, in most structural sections, be sufficiently slow that creep would eliminate any shrinkage stress. BS EN 1992, on the other hand, considers a completed piled raft as being exposed to end restraint and the reinforcement required to limit crack widths is significantly higher. End restraint was not even considered in BS 8007. We are not aware of significant problems with basements that were previously designed to that code.

BS EN 1992 calculates long-term shrinkage based on the average shrinkage through the depth of the concrete section. In a raft slab, drying can only occur from the top surface with the bottom surface often encased in a membrane and surrounded by water. Gilbert et al. (2012) measured the shrinkage profile through concrete that was sealed at the base as well as sealed and restrained at the base as in metal deck. They found a reduction in shrinkage of sections when evaporation was prevented from the base using a coating. However, when restrained and sealed, the base of the section exhibited no shrinkage. A similar effect would be expected to occur within piled rafts with shrinkage reduced at the base of the raft due to no evaporation from the bottom and restraint from distributed piles. Therefore the restrained drying shrinkage at the raft–pile interface may be much lower than anticipated from the average value calculated using CIRIA C660/BS EN 1992 and the reinforcement requirement significantly reduced. It can be very easy for the engineer to err on the side of caution and overdesign. It provides a greater factor of safety and it is not his money!

In these types of situations, an important impediment to more sustainable use of concrete has been a lack of in situ monitoring of structures to verify the design assumptions and minimise any overdesign. Advances in monitoring technology make it easier to acquire the required data. Accumulating more data on issues where overdesign may be occurring would also provide a technical basis for more sustainable construction without exposing the engineer to additional risk. We would urge the concrete industry to collect and publish as much data as possible on the in situ performance of concrete with appropriate cross-references to initial compliance testing.

13.1.4 Specifications

In the third edition of this book, Ken Day expressed the hope that the practice of specifying minimum cement contents and requiring mixes to be submitted and not subsequently varied would have finally died out by the publication of that edition of his book. However, these practices are still very much alive in 2012. Other prescriptive requirements of specifications, such as aggregate grading, maximum supplementary cementitious replacement levels, placement temperatures, and workability, tend to stifle mix optimisation and are an impediment to sustainability. They also often lead to unintended detrimental effects of concrete performance.

Designers of concrete structures and infrastructure should specify the properties they have assumed in their design, including strength, movement, and durability. However, few specifiers are also concrete technologists and many specifications are a blend of sometimes contradictory prescriptive and performance requirements. The performance requirements often just added onto previous specifications.

Existing codes accept that concrete strength follows a normal distribution and should be considered in terms of mean strength and standard deviation rather than an absolute limit. However, when cubes or cylinders are lower than specified strength, the engineer often requires an investigation by coring rather by analysing the results to determine whether the low result constitutes a genuine downturn or an isolated statistical aberration. To avoid the inconvenience and cost of coring or other testing, producers may choose to overdesign their concrete mixes, significantly reducing the sustainability of the concrete. Unnecessary testing of in situ concrete is an impediment to sustainability. Day has long advocated a penalty system where concrete that is "contractually" deficient results in a nominal cost to the producer. Clearly where compliance testing suggests "structurally" deficient concrete, an appropriate investigation would be required.

There are two basic requirements of a concrete control system. One should provide an accurate assessment of quality and the other should facilitate intervention as quickly as possible to restore the required quality in the event of any downturn. Accordingly, the specification must ensure that mix design and quality control are controlled by the concrete producer. Any external party cannot require corrective action based on as little evidence as a properly motivated producer will require. The large range of admixtures and supplementary cementitious materials now available makes external intervention even more difficult. A competent concrete producer has to conduct trials to establish which products, and which suppliers of materials, will best enable him to consistently produce the most economical compliant concrete for a particular project. He should be encouraged to do so by the specification. All parties to the project will benefit from a competent and motivated concrete supplier with consistent supply that complies with the specification requirements. Specifications for nonstrength properties can be more complicated and this is often used as a justification for prescribing some mix features, sometimes significantly reducing the sustainability of the concrete or the ability of the supplier to innovate.

Premature deterioration of reinforced concrete is a global problem that costs billions of dollars annually. In severe environments, concrete structures have often failed to achieve their required service life without major maintenance, which is unsustainable. As more specifications now require a minimum design life of 100 years or more for major projects and infrastructure, there is even more demand for appropriate specifications to ensure concrete durability. International codes provide prescriptive solutions to increase the required concrete quality and cover thickness to improve chloride resistance. The common practice to specify a minimum cementitious content to achieve "durability" is an impediment to sustainability. First, Buenfeld and Okundi (1998) showed that, at a given water to cementitious materials ratio (w/cm), the higher binder content actually increased chloride ion ingress in concrete. Similar results were found by Dhir et al. (2004). This is hardly surprising when transport processes occur primarily through the paste fraction of the concrete. Second, an unnecessarily high cementitious content may lead to increased cracking due to thermal stresses and shrinkage, which could reduce durability. Unnecessarily wasting cementitious materials also increases the environmental impact of the concrete. Another unintended consequence of minimum cementitious content requirements in specifications is that it creates a competitive disadvantage for the more competent concrete suppliers who have invested in effective quality control systems to be able to reduce variability and cementitious content.

One difficulty in specifying durability performance is the absence of a generally accepted comprehensive test at a reasonably early age. An increasing number of specifications require compliance testing of transport properties during construction in an attempt to improve the expected durability of reinforced concrete structures. However, the required performance for the different specified parameters to achieve the desired durability has often not been established. Unlike compressive strength, there is little information available on the expected variation in the results to calculate an appropriate characteristic value.

In the case of chloride-induced corrosion, performance requirements may include diffusion, migration, resistivity or water transport measurements, or combinations of these. The ASTM C1202 or Coulomb test has been a commonly specified procedure in different parts of the world. This procedure is a measurement of saturated resistivity and has been correlated to chloride diffusion. Although the standard includes a rough guideline for the interpretation of the coulomb values obtained, specifications often require more onerous performance limits, which appear more related to risk aversion than technical performance. Faced with onerous absolute performance limits suppliers have tended to significantly overdesign their concrete mixtures to help ensure compliance, which reduces sustainability and increases production cost with unknown benefit in terms of durability enhancement. The use of additional cementitious material to achieve certain performance limits at early ages may have a detrimental effect on fresh and hardened properties. The test result can have quite high variability so that individual results should not be specified as a rejection criterion for the sampled concrete, rather a characteristic value based on statistical analysis of results should be established.

Chloride diffusion is perhaps the most relevant test, but it is expensive and time consuming to test and therefore not well suited for compliance testing. Chloride migration is a much faster and cheaper procedure that still measures chloride penetration. As mentioned in Chapter 7, the improvement in chloride resistance of the concrete with time has been a difficult area in performance specification. We would suggest that the best procedure would be to conduct verification trials to determine the correlation between chloride migration and resistivity for the proposed mix and determine the improvement with time. For compliance testing, measure resistivity frequently and migration occasionally to confirm adequate performance based on service life predictions modeling using a characteristic value for assessment.

What is needed are more field data on the actual performance of concrete in aggressive environments related to its early-age properties to provide a better technical basis for performance requirements. There are good service life models that relate long-term field performance to early-age properties, but not all projects are going to conduct a detailed assessment of service life. However, simple and cheap compliance tests based on resistivity (for aggressive environments) and desorptivity (for water transport) could easily be added to compressive strength to provide much more information on the concrete's potential durability. When tests are cheap and simple, accumulating statistical data is easy and producers would be encouraged to get to understand how to optimise their mixes rather than the current situation of sticking to a mix because it has a diffusion coefficient or other expensive test data.

Specifications for temperature rise and differentials in massive pours require attention. A default peak temperature of 70°C is prudent as it would virtually eliminate the possible problem of delayed ettringite formation (DEF). Although DEF is uncommon, it can cause enormous damage.

Many specifiers focus on the temperature differential within the concrete mass and a value of 20°C is often specified. However, in our experience, the most significant thermal cracking has been caused by external restraint of massive concrete elements by a rigid substrate during cooling. The focus on the differential temperature requirement, particularly in temperate and tropical conditions, often leads to excessive insulation and increases both the peak temperature and the volume of concrete that reached high temperature. Therefore, to reduce a minor potential problem, the more likely a problem is often exacerbated.

Many specifications limit concrete placement temperature to 32°C or less. In hot countries, this usually means that premix companies need an ice plant and this has a high-energy demand. For a 4 metre thick concrete raft in Kuwait, the batching plant did not have an ice plant or access to flake ice but needed to achieve the required peak temperature limit. The use of 55% fly ash replacement achieved the required temperature limit as well as the other specified properties. In massive elements, very high replacement levels of fly ash and ground-granulated blast-furnace slag (GGBS) are extremely useful to limit temperature rise. The elevated temperature means that the in situ maturity is high at relatively early ages so that acceptable strength and penetrability properties do not take long to develop. There are many situations where in situ maturity monitoring can reduce unnecessary overdesign of concrete mixes. An unnecessary impediment to sustainability and solving potentially serious thermal issues are the limits on supplementary cementing material replacement levels in many specifications. One does need to be cautious when using high replacement levels of fly ash in thin or suspended elements where the concrete could dry out and not develop the required properties. Well-meaning "green" specifications that extend the compliance testing age for concrete specifically to enable high replacement levels without considering in situ strength development and other properties can be problematic.

There is a tendency to limit concrete workability in specifications based on the assumption that lower workability produces better concrete. Although often true when added water was the only way to increase workability, it is certainly not true in the age of advanced admixtures. Poor workability can lead to honeycombing, slower construction and uncontrolled water addition after compliance sampling. Resultant defects can lead to costly repairs and even litigation. The problem of prescriptive specification of rheology can also occur with self-consolidating concrete (SCC) where overzealous specifiers can require very high workability parameters, which can lead to segregation. I would suggest that the specification should require that the contractor or premix company confirm that the rheology of the concrete is satisfactory for the proposed placement procedure and the mix developed complies with the performance parameters. This will reduce the amount of repairs and replacement necessary and encourage innovation.

Many specifications include limits on drying shrinkage according to a standard procedure such as ASTM C157 or AS 1012.13. Although this may seem prudent and would be expected to reduce cracking, it should be noted that most "shrinkage" cracking is due to plastic, thermal, and autogenous shrinkage (in that order) not drying shrinkage. Drying shrinkage tests are conducted on well-cured small specimens 75 mm × 75 mm (3 inches × 3 inches) in cross-section dried at 50% relative humidity and therefore not representative of standard concrete elements exposed to drying in most environments. In situ drying shrinkage is a slow process. Pour strips interfere with construction and do virtually nothing to accommodate drying shrinkage strains in thicker slabs. Higher-strength concrete with higher cementitious contents tends to exhibit lower shrinkage in these tests. However, such mixes may have greater movement due to higher peak temperatures and more autogenous shrinkage, which are not measured in the test.

The well-meaning but poorly thought through use of a performance criterion may reduce both sustainability and concrete performance.

13.1.5 Regulations

Standard concrete production in Australia has essentially been based on compressive strength performance for more than 20 years. The Australian system resulted in good concrete producers, with well-equipped, suitably staffed, and accredited laboratories, designing and controlling a range of mixes to meet the specified strength. Concrete producers often prepared monthly reports on the mixes on the ConAd system, which were circulated to the purchasers. In the event of any marginally low result being predicted from early tests, the producer was expected to inform the purchaser of the concrete in question. The better producers generally use a graphical and statistical control system on concrete and input materials data, which helped identify any problem at an early stage.

The result of this system has been that typical concrete in Australia has a standard deviation of strength of between 2 and 3 MPa, well below most other countries. A lower standard deviation means a lower target mean strength and lower cementitious contents with reduced cost and environmental impact. Investment in quality control and quality testing was effectively incentivised.

The situation could not be more different for concrete from the same suppliers to various projects that must comply with prescriptive specification requirements, particularly when additional performance requirements have been added. Some state authorities have prescriptive specification requirements to which performance requirements have been added, such as chloride diffusion, sorptivity, or volume of permeable voids. Mixture proportions and material suppliers have to be registered with the state authority. The considerable time and cost involved in obtaining registration is a disincentive to ongoing mix development and upgrading to more advanced admixtures and so on. The standard deviation of these registered mixes with minimum cementitious content and other requirements can be up to double that of the standard mixes, which are controlled by the premix supplier to achieve a strength requirement only. When the variability in strength increases so does the variation in other properties.

An unexpected consequence of durability performance specification has been the submission of inappropriate concrete mixtures just because they had the necessary test data so the producer did not have to conduct additional trial mixes and long or expensive testing. For a structural element with minimum thickness varying from 0.45 m to 1.8 m and a specified strength of 50 MPa, the premix company proposed a mix with a cementitious content of 635 kg/m³ incorporating 25% fly ash. The mix was proposed because it had been approved by the appropriate statutory body and the performance criteria had been met. Modeling the proposed mix showed that he estimated peak temperature was 98°C (208°F) and the differential was more than 60°C (108°F). The mix may have achieved the required chloride diffusion but, if it had been used in this application, it would have resulted in severe cracking and significant delayed ettringite formation potential. In addition, the mix used over 200 kg/m³ more cementitious material than was necessary. In this situation, a performance requirement intended to improve durability and a registration procedure intended to ensure compliance could have resulted in the use of a totally inappropriate concrete mixture with serious consequences in terms of premature deterioration and waste of resources. This was simply to avoid additional testing and paperwork caused by the specification and regulations.

The Heart of Doha is an urban redevelopment in the historic center of the city. It will transform the district into a network of sustainable interconnecting buildings, public squares, courtyards, and landscaped streets. I was involved in helping improve the sustainable use of concrete for the project, which is targeting LEED™ Gold. Qatar had regulations preventing the establishment of a batching plant within the city—the "not in my backyard" rule. A comparison of the transport requirements for off-site compared with on-site concrete production of the estimated 1.25 million cubic metres showed a reduction of 55% in terms of truck kilometers for on-site production. Other important sustainability benefits were the ability to reduce loads on road infrastructure, reduce rejection, of noncompliant concrete, and reduce disruption to city traffic. Based on these benefits, permission was given to have a site plant.

13.1.6 Standards

Standards have necessarily been developed from the prevalent construction practices. Indeed, the time taken to develop standards means that they are often based on recent construction practices rather than current ones. This can be a serious impediment to the promotion and use of innovative materials and procedures. Although concrete standards are obviously based on Portland-cement-based concrete, if the materials components are essentially performance-based then innovative alternatives can be considered. Designers can request independent verification of the use of the product to help mitigate any possible risks with using a nontraditional concrete. This has been an excellent system for introducing innovative sustainable concrete materials in actual structures rather than laboratory specimens to build confidence in the technology.

National standards and codes which are more prescriptive in nature and explicitly limit concrete to a Portland-cement-based binder are an impediment to non-Portland-based binders being accepted in the industry.

13.1.7 Construction practices

Another impediment to the more sustainable use of concrete has been traditional construction practices that have covered concrete with marble, tiles, plaster, or paint. There is a different attitude to quality control of concrete when it is expected to have an off-form finish. We have seen repetitive defects where the first defect had been cosmetically patched rather than correcting the placement method that produced the defect. This has huge cost and sustainability implications. SCC can play an important role in improving the concrete quality and surface finish as well as saving contractors significant cost on repairs. The fact that so many precasters now use SCC is testament to its advantage in reducing defects and repairs.

Properly constructed concrete using appropriate binder, pigments, and formwork or grinding can provide an inexpensive, attractive, and durable finish where the thermal mass of the concrete is directly in contact with the internal spaces for maximum benefit in terms of thermal attenuation. This can be augmented by the use of embedded water pipes within the concrete to efficiently control internal temperatures.

13.1.8 Conclusions

We can produce beautiful, off-form structures with minimal embodied energy and emissions where most of its components are locally available throughout the world. Such concrete structures require virtually no maintenance. They are fire resistant, flood resistant, and hurricane proof in the event of severe weather with extremely low energy costs to maintain a comfortable living environment.

The concrete industry can play a huge role in sustainable development by eliminating excessive overdesign, rationalising specifications to promote quality and innovation, and improving construction practices. Some have become very cynical about sustainability because it can be more "spin" than substance and "spin" will not provide for the needs of humanity. However, if we all work to remove the various impediments to the sustainable use of concrete, this wonderful material can be used to its full potential in the service of mankind.

13.2 SUSTAINABILITY
Boudewijn M. Piscaer

The Pantheon in Rome demonstrates that concrete can be a highly sustainable material for a multifunctional building of almost 2000 years old. There is no such building in wood or metal. For fire resistance and now also for heat storage/exchange, concrete has an important role to play for sustainable living.

A "green" concrete that is not durable is not sustainable. Around 50% of durability failures in concrete are due to poor installation, 30% engineering design error and 20% poor mix. Most congresses and organisations focus on the 30 and 20% at an academic level. But do we include the installers and those who use the concrete?

Depending on the process roughly 1 ton ordinary Portland concrete (OPC) equals 1 ton of CO_2, which equals 1.6 tons of raw materials (600 kg lost as CO_2 into the atmosphere through calcination). In 2012 this amounts to 3.3 billion tons of CO_2 per year and 2 billion tons of raw materials lost by calcination. If we continue to produce concrete the same way as we do now, CO_2 from Portland cement could increase 260% by 2050 due to the increasing demand for concrete in the developing world. If most energy production will have been converted from fossil fuels to renewable sources, Portland cement could account for up to one-third of the global CO_2 output if we do not take action.

On average, 10 tons of aggregates are used per capita per year. Energy for production can vary from 3 kWh to 8 kWh per ton. Transport of aggregates has a considerable impact on society as well. In some mega cities, it has to come 200 km by truck. Availability is in some countries under pressure and this will result in using more nonconventional recycled materials.

From experience we know that more sustainable concrete does not have to cost more. Reducing the environmental impact of concrete goes hand in hand with improving social conditions, such as training plus increasing the prosperity of all involved in the industry and beyond. The primary task is to learn how to do more with less. Since Portland cement has the greatest environmental impact in concrete, the key objective of independent producers is to reduce its use. As we learned from high-temperature-resisting refractory concrete, when you use less of a critical ingredient, it has to be of higher quality. It can then provide a higher financial return for the supplier. More important, the people involved in the process at all levels have also to be better trained.

Since strength and durability do depend on reactivity, particle packing and adhesion especially of the fine particles, one has to learn to work with many different ingredients that are available at a reasonable cost and travel distance. Depending on the location, different supplementary cementing materials (SCMs) are available. Particle size engineering, especially of all the fines <125 μm with the objective to reduce the water demand of the mix, is becoming a key discipline.

SCMs play a key role in sustainability since most of them have much lower embodied energy and emissions. They also play different roles in reactivity, packing, and adhesion resulting in different strength development and durability. However, obtaining these attractive SCMs is often a local problem since they derive often as secondary products from other industries. New ones such as APReM (activated paper recycled minerals), rice husk ash (RHA), and SUCABM (sugar cane bagasse minerals) should be further industrially produced with proper quality control.

Also more choice of different fractions and better shapes of the aggregates will reduce the volume of the more expensive higher environmental impact paste.

Using the water/powder ratio as a guideline for mix design is replacing the water/cement ratio (w/c). The interpretation on what cement means is different from country to country and depends on different regulations. This has resulted in different calculated w/c or w/cm ratios depending on the treatment of different SCMs in different countries. Obviously such bureaucracy has little to do with technology.

13.2.1 Binders

Besides SCMs, both those recognised as ingredients of standardised composite blended cements and other more exotic ones, there are now many different new non-Portland clinker binders under development. Of all these, at the writing of the book, the family of alkali activated binders that include geopolymers seem to have gained the most credibility. The environmental impact of these binders versus clinker will depend on the sources of materials. Testing of materials that have a different rheology is being addressed by the RILEM.

13.2.2 From prescription to performance

Not how but what concrete has been produced is of importance. One could question what kind of automobiles we would have if the steel industry would have designed them. Is the steel content relevant to the performance of a car? Could the Pantheon in Rome be built using current prescriptive regulations? Especially in Europe there is a growing tendency to judge concrete on performance criteria including cost, workability, strength (at whatever age), durability, aesthetics, and now sustainability.

A "Sustainability Index—Concrete" became available in the Netherlands and Spain at the end of 2012. The Life Cycle Sustainability Assessment (LCSA) considers not only the technical aspects but also the social impact of products such as concrete. Attention to training is part of this index.

Having better nondestructive test equipment available will enhance the move from testing Lab-Crete toward Real-Crete performance. Remote monitoring of strength development at real temperatures as well as other properties can fine-tune the mix design and installation processes. In situ permeability testing, such as that developed by Dr Roberto Torrent, that is correlated to durability will improve workmanship that has to include good curing. And again, a green concrete that is not achieving the life it has been designed for is not sustainable.

Establishing standardisation for new products is mostly a costly time-consuming consensus seeking procedure that results often in mediocre compromises. Innovative mix designs will move toward a more dynamic verification according to a highly credible testing protocol in order to accelerate uptake and market confidence.

13.2.3 Participation

The awareness of what modern sustainable concrete means for people, the planet, and prosperity has to be improved. Early communication within the so-called participation pyramid—customer, specifiers (engineers and architects), suppliers, contractors, and the community—can dramatically improve the sustainability of concrete. At the end it is the sustainability of the concrete structure that counts.

For example, the aggregate for a bridge in Northern Norway was shipped over 1200 km, however, with a redesign local aggregates would have done the job. The real strength needed at a given time is more and more part of specifications that will influence the sustainability design of a mix.

Sustainable construction was a key issue in the selection of London for the 2012 Olympics. By promising to use sustainable concrete and other sustainable materials it actually saved around 10 million pounds. Early and continued communication within the participation pyramid done by a specialised company made this possible.

13.2.4 Opportunities

At the writing of this edition, the waste in concrete of CO_2, raw materials, energy, and human resources going to other disciplines is gigantic. A report from Harvard, "The Impact of a Corporate Culture of Sustainability on Corporate Behaviour and Performance", demonstrated that companies oriented toward sustainability have a higher return on investment by greater involvement of their people (Eccles et al., 2011). Sustainability in the concrete industry will have a highly positive impact on all stakeholders, though the move from volume toward quality thinking does not come easily for some.

- Concrete's image will be improved and better recognised as a high-tech material after having proven itself over 2000 years.
- Companies that coordinate sustainability of projects with all stakeholders will find value addition in the economy.
- Owners will be proud to use a safe, responsible, and affordable material.
- Architects and engineers will rediscover concrete with great new choices.
- From the producers of ingredients the cement industry will evolve from distributor toward sales engineers of high quality binders with good financial contributions.
- Aggregate producers will develop many different very precise and consistent fractions, including good quality dust, offering the lowest paste demand.
- Admixture producers will continue their good work as sales engineers involved in mix optimisation and maybe get more involved in mineral admixtures as well.
- Concrete producers will attract better people who get more satisfaction supplying a high-tech, tailor-made product for specific demand, than standard concrete.
- Research and technology companies that provide the producer with quality control (QC) of input and output materials, mix, and process optimisation will flourish.
- On-site inspection of Real-Crete using new nondestructive instruments for testing, for example, permeability will create new repair-cost-saving jobs.
- Sustainability evaluation companies that understand that managing is measuring will have more involvement with concrete to demonstrate its real impact.
- Vocational schools will graduate students with skills, proud to be involved with a high-tech material that is the most used globally and thus provide international mobility.
- Universities can present students, the industry, and public bodies with a range of new possibilities from research that may be applied more often in service of sustainable structures.

13.2.5 Water/cement and binder ratios

Globally there is still great confusion on what cement is. Are we talking about Portland cement/OPC or the 27 binders listed in the prescriptive EN 197? So how can we prescribe w/c ratios when the word *cement* itself is not clear? In Europe, when fly ash mixed with CEM I up to 35% by the cement producer, it is considered 100% cement and when mixed by the concrete producer it only counts for 40% as a binder. Is this what we need for our low-carbon-era concrete?

Looking at only the two mixes that both used 165 liters of water plus good PCE admixtures, would result in the following:

	WCR (EN197)	WBR NL	WBR ES+DE+	Water/powder ratio
SCC prec. C56/65	0.92	0.45	0.92	0.29
RMC C20/25	2.43	0.78	2.43	0.48

The water/powder ratio, not the w/c ratio or w/cm ratio explains the high strength.

13.2.6 Plan of action

Regulations

What kind of cars would we have if the steel industry had designed them? Do you think we would be able to build the Pantheon in Rome under present EN 197 and EN 206 national versions? Present regulations are restricting the use of low carbon binders by the concrete producer and discourage young people to work in this profession. On the other hand, we need credibility in the eyes of the customer so a verification methodology that allows a relative fast uptake of innovations seems to be the solution.

Although w/c ratio (or w/cm ratio) has done a job avoiding disasters in an uneducated setting, it would make sense to introduce the water/powder ratio as a guideline for concrete mix design only, where the powder component is all materials less than 125 microns. These examples highlight the fact that performance specification of properties must be allowed to take precedence over prescriptive requirements, which stifle innovation and the use of more environmentally friendly mixtures. Fortunately, more engineers and contractors are abandoning 28-day strength as mandatory in favor of environmental friendlier pozzolans such as fly ash.

As a first step, a producer should be free to use any SCM specified in the EN 197 to produce an "equal rights concrete" whether blended by the cement company or added separately at the batch plant. If the concrete producer is required to supply more tailor-made concrete, he should be allowed to tailor make binders.

Meanwhile the clinker producers can concentrate on making a CEM I 62.5 RR or CEM III C 22.5. For regular precasters they can supply the appropriate preblended binder. Second, those SCMs that are not recognised by EN 197 but have good reports should be considered, provided properties are verified. These include metakaolin, activated paper recycled minerals, reactive rue husk ash, and sugar cane bagasse minerals. Third, the scientific and technical community should actively follow the developments in non-Portland-based binders. After the recent conference on Durability in Trondheim, I became rather confident that alkali-activated binders have a future.

In the end it is not how we produce concrete according to suppliers but what concrete has been produced for the users.

Quality control and particle size engineered ingredients

A key element for low-carbon concrete is consistency of the ingredients that will allow precise design of both the structure and the mix. It will reduce waste due to usual overdesign of the mix to absorb variations. Improved incoming and outgoing QC will reduce the waste of valuable clinker. The technique of remote testing of strength development on the job will result in optimising mix and reduce the use of Portland cement.

Building capacity

Yet the most important element for sustainable concrete is building technical capacity at all levels, both academic and vocational. The world needs to know that concrete is a high-tech, durable product and that there is no structure in wood or metal that has lasted 2000 years. The people who work in this profession should be proud to be associated with the concrete industry.

Ground calcium carbonate

Ground calcium carbonate (GCC) has a very low carbon footprint. Using GCC to stabilise SCC mixes in the Netherlands in early 2000, nobody could make a C35 and C45 anymore. Maintaining the nationalised EN 206 version that prescribed the minimum amount of EN 197 cements, itself a prescriptive standard, they all increased concrete strength to C65 in 28 days, while having a demolding strength more than 20 MPa after 14 hours. This so-called inert material clearly contributed to strength, durability, and aesthetics bringing into question the link between w/c ratio and strength.

This SCC contained 180 kg/m^3 of GCC, GGBS, and CEM I. It would be considered to have a w/cm ratio of 0.40 in the United Kingdom; 0.45 in the Netherlands, 0.66 in France; and 0.92 in Germany, Spain, and many other countries. However, the same ingredients specified in the EN 197 at the concrete plant for making that SCC would have a w/cm ratio of 0.36.

In 2011 a pilot project was done in the framework of a European Eco-Innovation project SUSTCON EPV to develop a watertight basement in the Netherlands. Using only 68 kg/m³ CEM I 52.5 R (very fine, very stable OPC), 140 kg of GGBS and 132 kg GCC, a nominal 28-day strength of 25 MPa with good water resistance without additional special admixtures was obtained. This resulted in an estimated carbon footprint of less than 80 kg/m³ for the most commonly used concrete type (C 20/25) compared to the typical more than 250 kg/m³. Considering most countries would use more than 200 kg/m³ of CEM I/OPC, this pilot demonstrates that we can reduce CO_2 in concrete by more than half if we let go of outdated regulations.

13.3 MAGNESIUM-BASED CEMENTS

The third edition of the book, published in 2006, contained a section by John Harrison on magnesium oxide (MgO) in concrete, on which he had been working for many years. Currently, the addition of superfine calcium carbonate (limestone) replacing a portion of the cement content is becoming popular as a means of saving cement. Examining the resulting chemical balance, Harrison finds that, if substantial, this limits the amount of SCM (such as fly ash or slag), which can be effectively substituted for OPC. He suggests that the use of finely ground magnesium oxide instead of limestone would avoid this limitation. In addition, the MgO initially absorbs water, which it is able to later give up to enable more complete reaction of the cement and pozzolans.

In his worldwide search for funding and collaboration, one of his extensive contacts was with Imperial College in the United Kingdom. Imperial College obtained UK government funding, and started the group Novacem, to develop an alternative magnesium-based cement. We understand that it was considered to infringe on Harrison's patents and Novacem has been withdrawn from the market.

In our opinion, magnesium-based cements have an important future in reducing greenhouse gas and also having many desirable properties. Interested readers can directly access the work of John Harrison through this website: http://www.tececo.com.

13.4 IS GEOPOLYMER CONCRETE A SUITABLE ALTERNATIVE TO TRADITIONAL CONCRETE?

James Aldred and John Day

(Presented at Our World in Concrete & Structures in 2012.)
The term *geopolymer* was used by Davidovits (1991) to describe the inorganic aluminosilicate polymeric gel resulting from reaction of

amorphous aluminosilicates with alkali hydroxide and silicate solutions. Duxson et al. (2007) has identified many other names in the literature, such as alkali-activated cement, inorganic polymer concrete, and geocement, which have been used to describe materials synthesised using the same chemistry.

Synthesis of a geopolymer usually involves mixing materials containing aluminosilicates, such as metakaolin, fly ash, slag with alkali hydroxide, and alkali silicate solution, sometimes sodium carbonate in slag-based systems (Shi et al., 2006). There are numerous publications discussing different properties of geopolymers synthesised from different raw materials and activators. Therefore the term *geopolymer* covers a bewildering range of potential binders that those interested in this technology must navigate. Product data sheets, and even technical papers, on "geopolymers" might cherry pick data obtained from different binder chemistries giving the misleading impression that a specific material has been comprehensively tested when it has not. Papers might also focus on a particular material with poor performance to negatively characterise geopolymers. For example, the geopolymer concrete considered by Turner and Collins (2012) contained very high activator levels and required steam curing so that the product had relatively high embodied energy and emissions, leading to the conclusion that there was little benefit in terms of carbon footprint compared to ordinary Portland concrete (OPC).

One area where reference to generic geopolymer data is helpful is durability. For geopolymer concrete to be considered a suitable alternative to Portland-cement-based concrete, the basic geopolymeric gel must be durable. This can only be established over time. Xu et al. (2008) investigated activated slag concretes from the former Soviet Union. The slag had been activated by carbonates and by carbonate–hydroxide mixtures. The research found high compressive strengths that were significantly higher than when initially cast and excellent durability over a service life of up to 35 years. Xu et al. (2008) and Shi et al. (2006) report that the carbonation depths were relatively low for their age and no microcracks were observed after prolonged service. Although the performance of each proprietary geopolymer concrete needs to be established by comprehensive assessment, it is comforting to know that the basic geopolymer matrix appears to be durable and the reaction products appear stable over time.

Until recently, geopolymers have been found in niche applications, including fire-resistant materials, coatings, adhesives, and immobilisation of toxic waste[6] (Provis and van Derenter, 2009). However, the main potential application for geopolymers has been in the construction industry as an environmentally friendly concrete with reduced embodied energy and CO_2 footprint (Gartner, 2004; Phair, 2006) compared to the traditional Portland-cement-based concrete.

13.4.1 Mechanical properties

This geopolymer has been used on a number of different projects in Australia and a total volume of over 4000 m³ has been poured to date. It is not "labcrete"! Test specimens have been taken during actual production and a summary of the average mechanical properties are given in Table 13.1.

While the most common concrete grades used are 32 and 40 MPa (equivalent to fcu of 40 and 50 MPa), cylinder strengths up to 70 MPa have been measured. Since the geopolymer binder consists entirely of fly ash and ground-granulated blast-furnace slag (GGBS), there has been a common perception that geopolymer concrete would develop its strength very slowly or require heat curing. Portland cement systems containing high volume replacement of fly ash or GGBS and many geopolymer binders do develop compressive strength slowly. However, this particular geopolymer concrete develops its strength quite rapidly with design strength typically achieved after 7 days under laboratory conditions. Strength development at early age (up to 3 days) is sensitive to ambient temperature, but adequate early strength would be expected if the concrete temperature is above approximately 20°C.

The data available suggest that geopolymer concretes in general including this proprietary geopolymer tend to have higher tensile and flexural strength relative to the compressive strength than Portland-cement-based concrete. This appears due to the strong bond of the geopolymer gel to the aggregate particles (Concrete Institute of Australia 2011) and would be expected to improve crack resistance of geopolymer concrete.

Several researchers have reported a significantly lower elastic modulus for geopolymer concrete than for comparable OPC concrete. For example, Pan et al. (2011) found the reduction was about 23% for typical strength grade compared to the equations given in AS 3600. Accordingly those geopolymer concretes were outside guidelines given in Australian Standard 3600 and ACI Committee 363. However, the elastic modulus of this proprietary geopolymer concrete has been found to be comparable to Portland-cement-based concrete as shown in Table 13.1. The Poisson's ratio has been found to range between 0.19 and 0.24, which is slightly higher than would be expected for Portland-cement-based systems.

Table 13.1 Mechanical properties of geopolymer production concrete

Mix	Compressive strength (MPa)	Std. deviation	Tensile strength (MPa)	Flexural strength (MPa)	Shrinkage (microstrain)	Elastic modulus (GPa)	Poisson's ratio
32 MPa	38.1	3.7	4.5	6.2	300	31.8	0.20
40 MPa	55.6	4.3	6.0	6.6	230	38.5	0.24

13.4.2 Other significant properties

The drying shrinkage of this geopolymer concrete is much lower than for Portland-cement-based concrete with typical 56-day values of approximately 300 microstrain or less. The drying shrinkage will normally be less than that achieved for a Portland-cement-based concrete even incorporating a shrinkage reducing admixture as shown in Figure 13.1. The product also has a very low heat of hydration. The limited thermal and drying shrinkage makes it well-suited for thick and heavily restrained concrete elements and should enable a significant reduction in the quantity of crack control reinforcement.

While creep has not been directly measured, prestressed girders were cast using this proprietary geopolymer concrete in 2011. The prestress was transferred after 3 days. The girders were left unloaded for 100 days. The girders were loaded with W80 wheel load (8 tons) in accordance with the Australian bridge standard (AS 5100) and continuously measured for deflections over the subsequent 15-month period, as shown in Figure 13.2a. The hogging prior to load and deflection under sustained load were monitored using embedded vibrating wire strain gauges and the results are shown in Figure 13.2b. The structural behaviour in the girders was consistent with the compressive strength and modulus indicating no unusual deformation properties.

Precast reinforced beams were cast for the Global Change Institute at the University of Queensland. AECOM modeled the beam in RAPT based on an uncracked condition under self-weight and the measured mechanical properties. The expected deflection under the test load of 5 × 2 ton blocks equally spaced was calculated to be 3.0 mm. The actual maximum

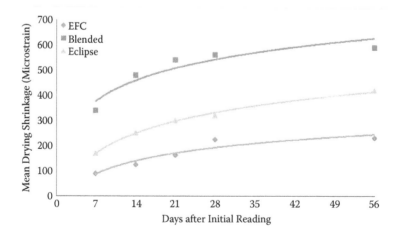

Figure 13.1 Drying shrinkage of geopolymer, 30% fly ash, and shrinkage reduced concretes.

(a)

(b)

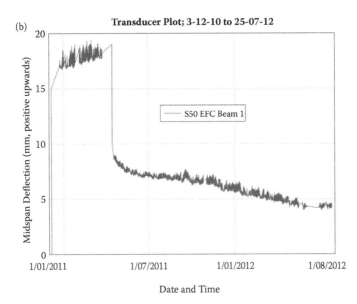

Figure 13.2 (a) Loading of prestressed geopolymer girder. (b) Initial hogging and loaded deflection of prestressed geopolymer girders.

deflection was 2.85 mm indicating that the structural behaviour of the beam closely followed the prediction.

The fire resistance of this proprietary geopolymer concrete has been tested according to the standard time-temperature curve (STTC) heating profile specified in the ISO 834 Standard for a cellulose fire. A structural panel (3 m × 4.7 m × 0.17 m) was installed into a specially designed furnace at the CSIRO Materials Science and Engineering Test Facility in Sydney. For the 2-hour test duration, it was exposed to a superimposed dead load of 5.5 kPa. The test showed this geopolymer concrete performed considerably better than would be expected for an OPC-based-concrete when exposed to the equivalent of a cellulose fire. The element satisfied the requirements of AS 1530 in spite of being exposed to full design load.

One point of concern that has been raised regarding the use of Portland cement free concrete is the potential for carbonation. Accelerated carbonation tests were conducted on a standard 40 MPa geopolymer concrete by RMIT. These tests showed that the depth of carbonation was higher than for an OPC concrete but was comparable to a 50 MPa concrete with 70% GGBS replacement.

The basic chemistry of this geopolymer concrete would be expected to provide good resistance to chloride and other aggressive chemicals. This has not been tested as yet. Samples were tested according to ASTM C1202 "Electrical Indication of Concrete's Ability to Resist Chloride Ion Penetration" and found to have "very low" chloride ion penetrability according to the guidelines (130–230 coulombs). RMIT have placed samples of this particular geopolymer at marine exposure sites in Fremantle, Portland, and Mackay, Australia, covering temperature and tropical exposure conditions.

13.4.3 Standards

Waste materials, such as fly ash and GGBS, are ideal to produce environmentally friendly geopolymer concrete. Fly ash and GGBS have been used with Portland cement in blended cement to reduce heat of hydration and improve other fresh and hardened properties. Their use in low heat cement application have been standardised for use by the Australian Standards Committee (Standards Australia, 2010) and International Standards Committee (ASTM International, 2010; British Standards Institution, 2011). The content of blended minerals usually vary greatly depending on the proposed use, for example, EN 197 CEM III/C cement allows up to 95% GGBS with 5% clinker. Many standards and specifications, such as EN 197, place limits on the alkali content of cement, fly ash, and GGBS, which, without qualification, may limit the acceptance of geopolymer-based products. As discussed by Shi et al. (2006), except in some former Soviet Union countries, there appear to have been no international standards or specifications for alkali-activated geopolymer concrete. In November 2010, the road regulator in the state of Victoria, Australia, (VicRoads) revised its specifications on "General Concrete Paving" (VicRoads, 2010). The introduction now states: "In the context of general concrete paving, portland cement concrete and geopolymer binder concrete are equivalent products".

This is a significant step for a major regulator in Australia and shows that VicRoads considers the data available on geopolymer concrete is sufficient to allow its use. The Concrete Institute of Australia published a "Current Practice Note" on geopolymer concrete in 2011, which may also help in the more widespread acceptance of this technology.

Standards have necessarily been developed from the prevalent construction practices. Indeed, the time taken to develop standards means that they are often based on recent construction practices rather than current ones. This can be a serious impediment to the promotion and use of innovative materials and procedures. James Aldred was involved in preparing a state-of-the-art report for this proprietary geopolymer concrete in Australia. While the standard is obviously based on Portland-cement-based concrete, the materials components of the AS 3600 for Concrete Structures are essentially performance based. The format of the report followed the engineering, durability, and other significant properties listed in the standard and compared the performance of the geopolymer concrete with the expected performance from a Portland-cement-based concrete. This approach has been quite successful in helping designers understand the performance properties of a novel material. This geopolymer concrete has now been used in a range of different applications. Designers have requested independent verification of the use of the product to help mitigate any possible risks with using a nontraditional concrete. This has been an excellent system for introducing innovative sustainable concrete materials in actual structures rather than laboratory specimens to build confidence in new technology.

National standards and codes that are more prescriptive in nature and explicitly limit concrete to a Portland-cement-based binder are an impediment to non-Portland-based binders being accepted in the industry. While SS 206-2009 (similar to EN 206) includes an equivalent performance concept, there is a restriction that potential binders should comply with EN 197 and therefore would technically exclude geopolymers that do not contain Portland cement clinker.

However, the BCA Green Mark System does strongly encourage the use of recycled materials and particularly innovation. Therefore there is good reason for Singaporean developers to look into this technology.

13.4.4 Field applications

Pavements

A typical light pavement 900 m long by 5.5 m wide was cast using grades 25 MPa and 40 MPa. A variety of construction procedures were used to assess pump compared with chute placement, saw cutting compared with wet-formed tooled joints, and manual compared with power troweling. A noticeable difference to Portland cement–based concrete is that

Figure 13.3 Placing of light pavement using geopolymer concrete.

the geopolymer concrete had no available bleed water rising to the surface. To maintain adequate surface moisture for screeding, floating, and
troweling operations as well as provide protection against drying, an
aliphatic-alcohol-based surface spray was used throughout the entire
placement period (Figure 13.3).

The pavement slab for weighbridge at the Port of Brisbane was cast in
November 2010 using grade 32 MPa geopolymer concrete. Geopolymer
has also been used in footpath applications by various local councils.

Retaining wall

A total of over fifty 40 MPa geopolymer precast panels were used a retaining wall for a private residence. The panels were up to 6 m long by 2.4 m
wide and were designed to retain earth pressure of 3 m. The precast panels
were cast in Toowoomba, Australia, and cured under ambient conditions
before being sent to site for installation (Figure 13.4).

Water tank

Two water tanks (10 m diameter × 2.4 m high) were cast in March 2011,
as seen in Figure 13.5. The first water tank was constructed using a grade
32 MPa concrete with a maximum aggregate size of 10 mm with blended
cement consisting of 80% Portland cement and 20% fly ash. The second
tank is constructed with a grade 32 MPa geopolymer concrete also with a
10 mm maximum aggregate.

Figure 13.4 Precast geopolymer retaining walls for a private residence.

Figure 13.5 In situ water tanks cast with blended cement concrete (left) and geopolymer concrete (right).

One reason was to investigate the autogenous healing behaviour of this geopolymer concrete. Autogenous healing in Portland-cement-based concrete is primarily due to the deposition of calcium hydroxide. As there is very little calcium hydroxide present in the geopolymer mix, the performance of geopolymer concrete in a water retaining application is of considerable interest.

Nominal leaking through cracks in the geopolymer tank did heal relatively rapidly. Ahn and Kishi (2010) suggest that geomaterials may be able to autogeneously heal due to a gel swelling mechanism.

Boat ramp

An extremely innovative application made possible under a research and development (R&D) project by QLD Transport and Main Roads, Department of Maritime Safety. The existing in situ concrete boat ramp at Rocky Point, Bundaberg, was due for replacement due to severe deterioration. Wagners were awarded an R&D tender to replace the ramp using an entirely novel form of construction material: precast concrete boat plank units made from grade 40 geopolymer concrete and reinforced with glass fibre reinforced polymer (GFRP) reinforcing bar. The approach slab on ground to the ramp was made from site cast geopolymer and similarly reinforced with GFRP. The project was successfully completed during November and December 2011, as seen in Figure 13.6. The precast ramp units were manufactured at Wagners precast facility in Toowoomba, while the site cast geopolymer for the approach slab was batched in Toowoomba, trucked to site with a 6.5-hour transit time, and then activated with the chemical activators on site. A unique feature of this particular geopolymer is that the entire batch constituents can be mixed in a truck bowl and remain completely dormant until the activator chemicals are added.

Precast bridge decks

One of the earliest fully structural applications of this geopolymer was the Murrarie Plant site bridge. This is a composite bridge structure made from pultruded fibreglass girders acting compositely with a grade 40 geopolymer bridge deck. The bridge was prefabricated at Wagners Toowoomba CFT factory and brought to site for installation in 2009, as seen in Figure 13.7. The bridge has been successfully in service since that date with continual concrete agitator truck loadings and no signs of distress.

The Bundaleer Road Bridge, West Moggill, Brisbane, was constructed and installed during May and June 2012, as shown in Figure 13.8. This project is another example of a composite pultruded girder and grade 40

Figure 13.6 Boat ramp constructed with both precast and in situ geopolymer concrete.

Figure 13.7 Installation of prefabricated bridge at Murrarie concrete batching plant.

geopolymer deck bridge structure. The geopolymer concrete deck acts as the compression flange to the bridge as well as providing a serviceable wearing deck. The client was the Brisbane City Council and the certifying engineer was i-cubed Pty Ltd.

Precast beams

The supply of grade 40 geopolymer to produce 33 precast floor beam-slab elements marks a significant milestone in modern geopolymer concrete. Believed to be the first application of modern geopolymer concrete into the structure of a multistory building, these precast floor beams form three suspended floor levels of the very innovative GCI building, which is a showcase for next-generation sustainable building technologies. There are two sizes of beams that span 10.8 m (× 2.4 m wide) and 9.6 m (× 2.4 m wide), respectively. Apart from being a structural floor element, the beams also are a major architectural feature, having an arched curved soffit and

Figure 13.8 Composite pultruded girder and grade 40 geopolymer deck bridge in Brisbane.

Figure 13.9 10.8 metre geopolymer beam with vaulted soffit being craned into position.

being specified as off-form class 2 with a light white colour, as seen being lifted into place in Figure 13.9. The beams will also play a major part in low-energy space heating with water pipes being placed inside them for temperature-controlled hydronic heating of the building spaces above and below. The building will be a showcase of sustainable construction. The project partners were: University of Queensland, principal; Hassell group, architect; Bligh Tanner, project engineer; AECOM, geopolymer certifying engineer; McNab, builder; and Precast Concrete Pty Ltd, precast manufacture.

13.4.5 Conclusions

Geopolymer binders cover a wide range of possible source materials and activators. Some binders within this generic group are not viable alternatives to traditional Portland-cement-based concrete. The particular geopolymer considered in this paper does appear to provide a suitable alternative and has been used in a number of applications in Australia. The low shrinkage and heat of hydration as well as the high tensile strength means that the material may have technical advantages over traditional concrete, particularly in structural elements subject to external restraint.

Standards and codes

AMERICAN SOCIETY FOR TESTING AND MATERIALS (ASTM)

ASTM C29/C29M–09. "Standard Test Method for Bulk Density (Unit Weight) and Voids in Aggregate".

ASTM C157/C157M–08. "Standard Test Method for Length Change of Hardened Hydraulic-Cement Mortar and Concrete".

ASTM C227–10. "Test Method for Potential Alkali Reactivity of Cement–Aggregate Combinations (Morta–Bar Method)".

ASTM C289–07. "Standard Test Method for Potential Alkali Silica Reactivity of Aggregates (Chemical Method)".

ASTM C294–12. "Standard Descriptive Nomenclature for Constituents of Concrete Aggregates".

ASTM C295/C295M–12. "Standard Guide for Petrographic Examination of Aggregates for Concrete".

ASTM C342–97. "Standard Test Method for Potential Volume Change of Cement Aggregate Combinations".

ASTM C441/C441M–11. "Standard Test Method for Effectiveness of Pozzolans or Ground Blast Furnace Slag in Preventing Excessive Expansion of Concrete Due to the Alkali Silica Reaction".

ASTM C457/C457M–12. "Standard Test Method for Microscopical Determination of Parameters of the Air Void System in Hardened Concrete".

ASTM C494–04. "Standard Specification for Chemical Admixtures for Concrete".

ASTM C586–11. "Standard Test Method for Potential Alkali Reactivity of Carbonate Rocks as Concrete Aggregates (Rock Cylinder Method)".

ASTM C618–12. "Standard Specification for Coal Fly Ash and Raw or Calcined Natural Pozzolan for Use in Concrete".

ASTM C642–13. "Standard Test Method for Density, Absorption, and Voids in Hardened Concrete".

ASTM C666/C666M–03(2008). "Standard Test Method for Resistance of Concrete to Rapid Freezing and Thawing".

ASTM C856–11. "Standard Practice for Petrographic Examination of Hardened Concrete".

ASTM C900–13. "Standard Test Method for Pullout Strength of Hardened Concrete".

ASTM C1074–11. "Standard Practice for Estimating Concrete Strength by the Maturity Method".

ASTM C1202–97. "Standard Test Method for Electrical Indication of Concrete's Ability to Resist Chloride Ion Penetration".

ASTM C1252–06. "Standard Test Methods for Uncompacted Void Content of Fine Aggregate (as Influenced by Particle Shape, Surface Texture, and Grading)".

ASTM C1260–07. "Standard Test Method for Potential Alkali Reactivity of Aggregates (Mortar-Bar Method)".

ASTM C1293–08. "Standard Test Method for Determination of Length Change of Concrete Due to Alkali Silica Reaction".

ASTM C1556–04. "Standard Test Method for Determining the Apparent Chloride Diffusion Coefficient of Cementitious Mixtures by Bulk Diffusion".

ASTM C1567–13. "Standard Test Method for Determining the Potential Alkali–Silica Reactivity of Combinations of Cementitious Materials and Aggregate (Accelerated Mortar-Bar Method)".

ASTM C1611 / C1611M–09. "Standard Test Method for Slump Flow of Self Consolidating Concrete".

ASTM G109–07(2013). "Standard Test Method for Determining Effects of Chemical Admixtures on Corrosion of Embedded Steel Reinforcement in Concrete Exposed to Chloride Environments".

STANDARDS AUSTRALIA (AS)

AS 1012.3.3–98. "Methods of Testing Concrete—Determination of Properties Related to the Consistency of Concrete—Vebe Test".

AS 1012.13–92. "Methods of Testing Concrete—Determination of the Drying Shrinkage of Concrete for Samples Prepared in the Field or in the Laboratory".

AS 1012.21–99. "Methods of Testing Concrete—Determination of Water Absorption and Apparent Volume of Permeable Voids in Hardened Concrete".

AS 1530.1–94. "Methods for Fire Tests on Building Materials, Components, and Structures—Combustibility Test for Materials".

AS 3582.1–98. "Supplementary Cementitious Materials for Use with Portland and Blended Cement—Fly Ash".

AS 3582.2–01. "Supplementary Cementitious Materials for Use with Portland and Blended Cement—Slag—Ground Granulated Iron Blast-Furnace".

AS/NZS 3582.3–02. "Supplementary Cementitious Materials for Use with Portland and Blended Cement—Amorphous Silica".

AS 3600–09. "Concrete Structures".

AS 4997–05. "Guidelines for the Design of Maritime Structures".

AS 5100.1–04. "Bridge design—Scope and General Principles".

BRITISH AND EUROPEAN STANDARDS (BS EN)

BS 1881: Pt 122: 1983. "Method for Determination of Water Absorption". British Standards Institute, London.

BS 1881: Pt 208: 1996. "Test for Determining the Initial Surface Absorption of Concrete". British Standards Institute, London.

BS 6349-1-3: 2012. "Maritime works. General. Code of Practice for Geotechnical Design".

BS 8007: 1987. "Code of Practice for Design of Concrete Structures for Retaining Aqueous Liquids".

BS 8110-1: 1997. "Structural Use of Concrete. Code of Practice for Design and Construction".

BS 8110-2: 1985. "Structural Use of Concrete. Code of Practice for Special Circumstances".

BS 8500-1: 2002. "Concrete. Complementary British Standard to BS EN 206-1. Method of specifying and guidance for the specifier".

BS EN 197-1: 2011. "Cement. Composition, Specifications, and Conformity Criteria for Common Cements".

BS EN 206-1: 2000. "Concrete. Specification, Performance, Production, and Conformity".

BS EN 934-2: 2001. "Admixtures for Concrete, Mortar, and Grout. Concrete Admixtures. Definitions, Requirements, Conformity, Marking, and Labeling".

BS EN 1992-1-1: 2004. "Design of Concrete Structures. General Rules and Rules for Buildings".

BS EN 12390-8: 2009. "Testing Hardened Concrete. Depth of Penetration of Water Under Pressure".

AMERICAN CONCRETE INSTITUTE (ACI)

ACI 209.2R-08. "Guide for Modeling and Calculating Shrinkage and Creep in Hardened Concrete".

ACI 211.4R-93. "Guide for Selecting Proportions for High-Strength Concrete with Portland Cement and Fly Ash".

ACI 212.3R-04. "Chemical admixtures for concrete". America Concrete Institute, Farmington Hills, MI, 2004.

ACI 225R-99. "Guide to the Selection and Use of Hydraulic Cements". (Reapproved 2009.)

ACI 232.1R-12. "Report on the Use of Raw or Processed Natural Pozzolans in Concrete".

ACI 305.1-06. "Specification for Hot Weather Concreting".

ACI 318-11. "Building Code Requirements for Structural Concrete and Commentary".

OTHER STANDARDS AND CODES

AASHTO T 318–02 (2011). "Standard Method of Test for Water Content of Freshly Mixed Concrete Using Microwave Oven Drying".

Nordtest NT Build 443: 1995. "Concrete, Hardened: Accelerated Chloride Penetration".

Nordtest NT Build 492: 1999. "Concrete, Mortar, and Cement-Based Repair Materials: Chloride Migration Coefficient from Non-Steady-State Migration Experiments".

SS EN 206–1: 2009. "Concrete—Specification, Performance, Production, and Conformity".

References

Abrams, D.A. 1927. Water–cement ration as a basis of concrete quality. *Journal Proceedings*, 23, 2, 452–457.

Ahn, T.-H., and Kishi, T. 2010. Crack self-healing behavior of cementitious composites incorporating various mineral admixtures. *Journal of Advanced Concrete Technology* , 8, 2, 171–186.

Aitcin, P.C. 2011. *High Performance Concrete*. Boca Raton: Taylor & Francis.

Aldred, J.M. 1988. Sorptivity as an indicator of concrete durability in marine environments. Suppl. Papers of 2nd International Conference on Performance of Concrete in Marine Environment. St Andrews By-the-Sea, Canada: ACI-CANMET.

Aldred, J.M. 1989. The short-term and long-term performance of concrete incorporating dampproofing admixtures. In Supplementary Papers of 3rd CANMET/ACI International Conference on Superplasticizers and Other Chemicals in Concrete, Oct. 1989, Ottawa, Canada, 1–19. United States: ACI.

Aldred, J.M. 1999. Chloride and water movement in concrete with and without admixtures. M. Eng thesis. Singapore: National University of Singapore.

Aldred, J.M. 2008. Water transport due to wick action through concrete. Doctor of Philosophy. Kent, Australia: Department of Civil Engineering, Curtin University of Technology.

Aldred, J.M., and Lim, S.N. 2004. Factors affecting the autogenous shrinkage of ground granulated blast-furnace slag concrete. 8th CANMET/ACI International Conference on Fly Ash, Silica Fume, Slag and Natural Pozzolans in Concrete. ACI SP 221 pp. 783–796.

Aldred, J.M., Swaddiwudhipong, S, Lee, S.L., and Wee, T.H. 2001. The effect of initial moisture content on water transport in concrete containing a hydrophobic admixture. *Magazine of Concrete Research*, 53, 2, 127–134.

American Concrete Institute. 1998. Guide for selecting proportions for high-strength concrete with Portland cement and fly ash. ACI 211.4R–93.

Andrews-Phaedonos, F. 2012. Testing concrete durability—Need to use a single parameter with high level of accuracy for mix design, quality control, and in-situ assessment. 25th ARRB Conference, Perth, Australia.

Ann, K.Y., and Song, H-W. 2007. Chloride threshold level for corrosion of steel in concrete. *Corrosion Science*, 49, 4113–4133.

Ann, K.Y., Jung, H.S., Kim, H.S., Kim, S.S., and Moon, H.Y. 2006. Effect of calcium nitrite-based corrosion inhibitor in preventing corrosion of embedded steel in concrete. *Cement and Concrete Research*, 36, 530–535.

ASTM International. 2010. ASTM C595/C595M—10 standard specification for blended hydraulic cements. ASTM, 04.01, September.

Arrhenius, S. 1896. On the influence of carbonic acid in the air upon the temperature of the ground. *Philosophical Magazine and Journal of Science Series*, 5, 41, 237–276.

Audenaert, K., Yuan, Q. and De Schutter, G. 2010. On the time dependency of the chloride migration coefficient in concrete. *Construction and Building Materials*, 24(3), 396–402.

Bamforth, P.B. 1980. In situ measurement of the effect of partial Portland cement replacement using either fly ash or GGBS on the performance of mass concrete. In *Proceedings of the Institution of Civil Engineers*, 2, 69, 777–800.

Bamforth, P.B. 2007. Early-age thermal crack control in concrete. CIRIA C660.

Bamforth, P.B. and Price, W.F. 1993. Factors influencing chloride ingress into marine structures. Proceedings of Concrete 2000, Dundee. London: E & FN Spon, 1105–1118.

Baroghel-Bouny, V., Mainguy, M. and Coussy, O. 2001. Isothermal drying process in weakly permeable cementitious materials – assessment of water permeability. In *Materials Science of Concrete: Ion and Mass Transport in Cement-Based Materials* (R.D. Hooten, M.D.A. Thomas, J. Marchand, and J.J. Beaudoin, eds.). Special Volume, American Ceramic Society, October.

Bentz, D.P, and Hansen, K.K. 2000. Preliminary observations of water movement in cement pastes during curing using X-ray absorption. *Cement & Concrete Research* V.30, 1157–1168.

Bentz, D.P, Geiker, M.R., and Hansen, K.K. 2001. Shrinkage-reducing admixtures and early-age desiccation in cement pastes and mortars. *Cement and Concrete Research*, 31, 1075–1085.

Bertrandy, R. 1982. Sands used in construction concrete—The influence of fines on their properties. Unpublished paper translated and presented by K. Day. Singapore, November.

Bilodeau, A., and Malhotra, V.M. 1994. High-performance concrete incorporating large volumes of ASTM Class F fly ash. ACI Special Publication, vol. 149–10.

Blick, R. L., Petersen, C. F., and Winter, M. E. 1974. Proportioning and Controlling High Strength Concrete, Proportioning Concrete Mixes, AC! Special Publication 46–9.

Bolomey, J. 1947. The grading of aggregate and its influence on the characteristics of concrete. *Revue Mater Construction, Travaux Publiques*.

Bouley, C., and de Larrard, F. 1993. The sand box test. Concrete International: Design and Construction, April.

BRE Special Digest 1. 2005. *Concrete in aggressive ground,* 3rd edition. BRE Bookshop.

Buenfeld, N.R., and Okundi, E. 1998. Effect of cement content on transport in concrete. *Magazine of Concrete Research*, 50, 4, 339–351.

Buenfeld, N.R., Shurafa-Daoudi, M.T., and McLoughlin, I.M. 1997. Chloride transport due to wick action in concrete. In Chloride Penetration into Concrete (L.O. Nilsson and J. P. Ollivier, eds.). RILEM, Paris, pp. 315–324.

Buhler, E.R. 2007. Two decades of ready-mixed high performance silica fume Concrete—A U.S. project review. Dallas, TX: NRMCA Presentation.

Building Research Advisory Board. 1958. Effectiveness of concrete admixtures in controlling the transmission of moisture through slabs-on-ground. Publication No. 596. Washington, DC: N. R. C.

Bungey, J.H. 1993. Non-Destructive Testing in Civil Engineering.

Butler, W.B. 1994. Superfine fly ash in high strength concrete. Dundee, Scotland: Concrete 2000.

Calder, A.J.J., and Thomson, D.M. 1988. Repair of cracked reinforced concrete: assessment of corrosion protection. TRL – RR150. United Kingdom.

Cao, T. H.T., Meck, E. and Morris, H. 1996. A review of the ASTM C1202 Standard Test and its applicability in the assessment of concrete's resistance to chloride ion penetration. Concrete in Australia, 23–36.

Carino, N. J. 1984. Maturity method: Theory and application, cement, concrete, and aggregates. ASTM, 6, 2.

Carrasquillo, P.M., and Carrasquillo, R.L. 1988. Effect of using unbonded capping system on the compressive strength of concrete cylinders. ACI Materials Journal, 85(3), 141–147.

Cement: Composition, specifications, and conformity criteria for low heat common cements. 2000. BS EN, 197–1.

Chung, H. W. 1978. How good is good enough? A dilemma in acceptance testing of concrete. ACI Journal, Proceedings, 75, 8, 374–380, August.

Clelland, J. 1968. Sand for concrete—A new test method. New Zealand Standards Bulletin, 22–26.

Concrete Institute of Australia. 2011. Recommended Practice—Geopolymer Concrete Z16.

Concrete Society CS-163. 2008. Guide to the design of concrete structures in the Arabian Peninsula.

Concrete Society. 2008. Permeability testing of site concrete—A review of methods and experience. Technical Report 31.

Cordon, W.A., and Merrill, D. 1963. Requirements for freezing and thawing durability for concrete. Proceedings of ASTM, 63, 1026–1036.

Dalhuisen, P. et al. 1996. Fourth International Symposium on Utilisation of High Strength/High Performance Concrete. Paris: Presses Ponts et Chausees.

Davidovits, J. 1991. Geopolymers: Inorganic polymeric new materials. Journal of Thermal Analysis, 37, 1633–1656.

Day, K. W. 1959. Mathematical methods of proportioning aggregates. Concrete and Constructional Engineering. London, February.

Day, K. W. 1961. The Specification of concrete. Constructional Review (Australia), 34, 7, 45–58, July.

Day, K. W. 1981. Quality control of 55 MPa concrete for Collins Place Project, Melbourne, Australia. Concrete International: Design and Construction, 3, 3, 1724. (Presented at ACI Milwaukee Convention, March 1979.)

Day, K. W. 1982a. What is economical concrete? Concrete International: Design and Construction, September.

Day, K. W. 1989. Bad concrete or bad testing? Unpublished paper to ACT San Diego Convention, November.

Day, K.W. 1987. Marginal sands. San Antonio, TX: ACI Convention.

Day, K. W. 1996a. Production of high performance concrete. Fourth International Symposium on Utilization of High Strength/High Performance Concrete, May. Paris: Presses Ponts at Chausses.

Day, K. W. 2002. Just-in-time mixture proportioning. Cancun, Mexico: ACI Convention, December [see author's website].

Day, K. W. 2005. Concrete in the 22nd century. Concrete Institute of Australia Biennial Symposium, Melbourne, Australia, October [see author's website].

Day, K.W. 2008. Optimised concrete quality control. Holland: FIB Conference.

Day, R.L. 1992. Pozzolans for use in low-cost housings: A state of the art report. Research Report No. CE92-1. Calgary, Canada: Department of Civil Engineering, University of Calgary.

de Champs, J. F., and Monachon, P. 1992. Une application remarkable: L'arc du pont stir la Rance, Les betons a haut performances—characterization, durabilite, applications. Paris: Presses des Pontes at Chaussees.

de Larrard, E. (ed.). 1999. *Concrete Mixture Proportioning: A Scientific Approach.* London: E & FN Spon.

Dewar, J. D. 1999. *Computer Modelling of Concrete Mixtures.* London: E & FN Spon.

Dhir, R.K, Hewlett, P.C., and Chan, Y.N. 1987. Near-surface characteristics of concrete: assessment and development of in situ test methods. *Magazine of Concrete Research,* 39, 141.

Dhir, R.K., McCarthy, M.J., Zhou, S., and Tittle, P.A.J. 2004. Role of cement content in specifications for concrete durability: Cement type influences. *Structures & Buildings,* 157(SB2), 113–127.

Diamond, S. 1971. A critical comparison of mercury porosimetry and capillary condensation pore size distributions of portland cement pastes. *Cement and Concrete Research,* 1, 531–545.

Diamond, S. 2000. Mercury porosimetry—An inappropriate method for the measurement of pore size distributions in cement-based materials. *Cement and Concrete Research,* 30, 1517–1525.

Dolch, W., and Lovell, J. 1988. Wetting and drying as indicators of water/cement ratio. Conference on Concrete Durability, ACI SP-100, 509–517.

Duxson, P., Fernández-Jiménez, A., Provis, J.L., Lukey, G.C., Palomo, A., and van Deventer, J. 2007. Geopolymer technology: The current state of the art. *Journal of Materials Science,* 42, 2917–2933.

Eccles, R.G., Ioannou, I. and Serafeim, G. 2011. The impact of corporate sustainability on organizational processes and performance. Harvard Business School Working Paper Series 12-035. Available at SSRN, http://ssrn.com/abstract=1964011 or http://dx.doi.org/10.2139/ssrn.1964011.

Elek, A. 1973. A test for assessment of fine aggregates. *Hume News,* 11–12.

Erdogan, S.T., and Fowler, D.W. 2005. Determination of aggregate shape properties using x-ray tomographic methods and the effect of shape on concrete rheology. Report No. ICAR 106-1.

Espinosa, R.M., and Franke, L. 2006a. Inkbottle pore-method: Prediction of hygroscopic water content in hardened cement paste at variable climatic conditions. *Cement and Concrete Research,* 36, 1956–1970.

Espinosa, R.M., and Franke, L. 2006b. Influence of the age and drying process on pore structure and sorption isotherms of hardened cement paste. *Cement and Concrete Research,* 36, 1971–1986.

Farny, J.A. and Kerkhoff, B. 2007. *Diagnosis and Control of Alkali-Aggregate Reactions in Concrete*. IS413. Skokie, IL: Portland Cement Association.

Faury, J. 1958. *Le Beton, 3rd edition*.

Feret. 1896.

FHWA. 2006. Freeze–thaw resistance of concrete with marginal air content. Publication No. FHWA-HRT-06-117.

Foo, K.Y., and Hameed, B.H. 2009. Utilization of rice husk ash as novel adsorbent: A judicious recycling of the colloidal agricultural waste. *Advances in Colloid and Interface Science*, 152, 39–47.

Gartner, E. 2004. Industrially interesting approaches to "low-CO2" cements. *Cement and Concrete Research*, 34, 1489–1498.

Gaynor, R.D. 1967 and 1968. Exploratory tests of concrete sands. JRL Series 190 Report. Silver Spring, MO: National Sand & Gravel Association/National Ready Mixed Concrete Association.

Gilbert, R.I., Bradford, M.A., Gholamhoseini, A., and Chang, Z.-T. 2012. Effects of shrinkage on the long-term stresses and deformations of composite concrete slabs. *Engineering Structures*, 40, 9–19.

Graham, G., and Martin, F. R. 1946. The construction of high-grade quality concrete paving for modern transport aircraft. *Journal of Institution of Civil Engineers*, 26, 6, 117–190, April.

Guo, C. 1989. Maturity of concrete: Method for prediction of early age strength. *ACI Materials Journal*, July–August (and discussion May–June 1990).

Hammond, A.A. 1983. Pozzolana cements for low cost housing. Appropriate Building Materials for Low Cost Housing. Proceedings of Symposium, Nairobi, Kenya. New York: E & F.N. Spon, pp. 73–83.

Hansen, J., Fung, I., Lacis, A., Rind, D., Lebedeff, S., Ruedy, R., Russell, G., and Stone, P. 1988. Global climate changes as forecast by Goddard Institute for Space Studies—three dimensional model. *Journal of Geophysical Research*, 93, D8, 9341–9364.

Harrison, N.L. 1988. Description and assessment of sands. Humes report, RC.1562.

Hearn, N. 1998. Self-sealing, autogenous healing and continued hydration: What is the difference? *Materials and Structures*, 31. 563–567.

Ho, D.W.S., and Lewis, R.K. 1984. Concrete quality as measured by water sorptivity. Transactions of the Institution of Engineers of Australia, Civil Engineering Transactions, CE26, 306–313.

Ho, D.W.S., and Lewis, R.K. 1988. The specification of concrete for reinforcement protection – Performance criteria and compliance by strength. *Cement and Concrete Research*, 18, 584–594.

Hopkins, H.J. 1971. Sands for concrete—A study of shapes and sizes. *New Zealand Engineering*, 287–292.

Hughes, B. R. 1954. Rational concrete mix design. *ICE Journal*.

Jennings, H.M. 2000. A model for the microstructure of calcium silicate hydrate in cement paste. *Cement and Concrete Research*, 36, 101–116.

Kaplan, D., de Larrard, F., and Sedran, T. 2005. Avoidance of blockages in concrete pumping process. *ACI Materials Journal*, 102-M21, 183–191.

Kaplan, M. E. 1958. The effects of the properties of coarse aggregates on the workability of concrete. *Magazine of Concrete Research*, 29, August, 63–74.

Khatri, R. P., Hirschausen, D., and Sirivivatnanon, V. 1998. Broadening the use of fly ash concretes within current specifications. CSIRO Report BRE045.

Kerrigan, B.M. 1972. Sand flow test. Humes report, RC.4243.

Kluge, F. 1949. Prediction of the mixing water of concrete. *Der Bauingenuier*, 24, 6. [In German.]

Lagerblad, B., and Utkin, P. 1993. Silica granules in concrete—dispersion and durability aspects. Swedish Cement and Concrete Research Institute. CBI Report 3:93.

Leshchinsky, A. 2004. Slag sand in ready-mixed concrete. *Concrete*, 38(3), 38–39.

Lessard, M., Chaallal, O., and Aitcin, P.-C. 1993. Testing high-strength concrete compressive strength. *ACI Materials Journal*, 90(4), 303–308.

Leviant, I. 1966. A graphical method of concrete proportioning. In *Civil Engineering and Public Works Review*. London: Lomax, Erskine, and Company.

Malier, Y., and Moranville-Regourd, M. 1995. High performance concrete, the French national project. Durability of High Performance Concrete (H. Sommer, ed.). Paris: RILEM, 78–88.

Malhotra, V.M. 1964. Correlation between particle shape and surface texture of fine aggregate and their water requirement. *Materials Research & Standards*, 656–658.

Malhotra, V.M., and Ramezanianpour, A.A. 1994. *Fly Ash in Concrete,* 2nd edition. CANMET.

Marchand, J., Bentz, D., Samson, E., and Maltais, Y. 2001. Influence of calcium hydroxide dissolution on the transport properties of hydrated cement systems. In *Calcium Hydroxide in Concrete* (J. Skalny, J. Gebauer, and I. Odler, eds.). Westerville, OH: American Ceramic Society.

Maslehuddin, M., Shameem, M., Ibrahim, M., and Khan N.U. 1999. Performance of steel aggregate concretes, exploiting waste in concrete. *Proceedings of International Conference Creating with Concrete*. Dundee, Scotland: Thomas Telford Ltd., pp. 109–119.

Mather, B. 1987. The warmer the concrete, the faster the cement hydrates. Concrete International: Design and Construction, August.

Mehta, P.K. 1987. Natural pozzolans. In Supplementary Cementing Materials for Concrete (V.M. Malhotra, ed.). CANMET-SP-86-8E. Ottawa, Canada: Canadian Government, pp. 1–33.

Mielenz, R. C., Greene, K. T., and Schieltz, N. C. 1951. Natural pozzolans for concrete. *Economic Geology*, 46, 3, 311–328.

Mitsuki, Y.et al, 1992. An enhancement in the nature of concrete with a multiplicative cement crystal type concrete material. 46th Annual Meeting of the Civil Engineering Society, September.

Mohammed, T.U., and Hamada, H. 2003. Corrosion of steel bars in concrete at joints under tidal environment. ACI Materials Journal Title no. 100-M31, 265–273.

Montes, P., Bremner, T.W., and Lister, D.H. 2004. Influence of calcium nitrite inhibitor and crack width on corrosion of steel in high performance concrete subjected to a simulated marine environment. *Cement & Concrete Composites*, 26, 243–253.

Murdock, L. J. 1960. The workability of concrete. *Magazine of Concrete Research*, 12, 36, 135–144.

Neville, A.M. 1981. *Properties of Concrete, 3rd Edition*. Pitman.

Neville, A.M. 1987. Why do we have concrete durability problems? Katharine and Bryant Mather Conference on Concrete Durability, ACI SP-100, 1.

Neville, A.M. 2011. *Properties of concrete*, 5th ed. Harlow, England: Pearson.

Nokken, M.R. 2004. Development of discontinuous capillary porosity in concrete and its influence on durability. PhD thesis. Toronto, Canada: Department of Civil Engineering, University of Toronto.

Nurse, R. W. 1949. Steam curing of concrete. *Magazine of Concrete Research*, 1, 2.

Ozyildirim, C. 2011. Virginia's end-result specifications. Concrete International, March, pp. 41045

Odler, I. 1991. High volume fly ash.

Pan, Z., Sanjayan, J.G., Rangan, B.V. 2011. Fracture properties of geopolymer paste and concrete. *Magazine of Concrete Research*, October.

Odler, I. 1991. Final report of Task Group 1, 68—MMH technical committee on strength of cement. *Materials and Structures*, 24, 140, 143–157.

Parrott, L.J. 1991. Factors influencing relative humidity in concrete. *Magazine of Concrete Research*, 43, 154, 45–52.

Petinau, C.B. 2003. The effects of the type and quantity of binder on the adiabatic temperature rise in mass concrete. Final year project. Kent, Australia: Curtin University of Technology.

Phair, W. 2006. Green chemistry for sustainable cement production and use. *Green Chemistry*, 8, 763–780.

Pocock, D., and Corrans, J. 2007. Concrete durability testing in the Middle East. *Concrete Engineering International*, Summer, 52–54.

Popovics, S. 1982. *Fundamentals of Portland Cement Concrete, Vol. 1: Fresh Concrete*. Hoboken, NJ: John Wiley.

Popovics, S. 1992. *Concrete Materials, Properties, Specifications, and Testing*. Noyes.

Powers, L.J. 1999. Developments in alkali-silica gel detection. *Concrete Technology Today*. PL991. Portland Cement Association. Accessed from http://www.cement.org/pdf_files/PL991.pdf.

Powers, T.C., Copeland, L.E., Hayes, J.C., and Mann, H.M. 1954. Permeability of Portland cement paste. *ACI Proceedings*, 51, 285–298.

Prat. 1996.

Provis, J.L., and van Deventer, J.S.J. (eds.). 2009. *Geopolymers: Structures, Processing, Properties, and Industrial Applications*. Cambridge, UK: Woodhead Publishing Limited.

RILEM. 1995. Performance criteria for concrete durability (J. Kropp, and H. K. Hilsdorf, eds.). RILEM report 12. London: E & F.N. Spon.

Rixom, M.R., and Mailvaganam, N.P. 1986. *Chemical Admixtures for Concrete*. London: E & F.N. Spon.

Robbery, P.C. 1987. Waterproofers. Admixtures for Concrete Construction. Concrete Society, September.

Roberts, M . H., and Adderson, B. W. 1985. Tests on water resisting admixtures for concrete. B.R.E Technical Note No. 159/85.

Roy S.K., Northwood D.O., and Aldred, J.M. 1995. Relative effectiveness of different admixtures to prevent water penetration in concrete. ConChem: International Exhibition and Conference, Brussels, 451– 459.

Sagues, A.A., Powers, R.G., and Kessler, R. 2001. Corrosion performance of epoxy-coated rebar in florida keys bridges. Paper 01642 NACE International Conference, Houston, Texas.

Samson, E. and Marchand, J. 1999. Numerical solution of the extended Nernst–Planck model. *Journal of Colloid and Interface Science*, 215, 1–8.

Saul, A. G. A. 1951. Principles underlying the steam curing of concrete at atmospheric pressure. *Magazine of Concrete Research*, 2, 6, 127–440.

Sherman, M.R., McDonald, D.B., and Pfeifer, D.W. 1996. Durability aspects of precast prestressed concrete, part 2: Chloride permeability study. *Precast Concrete Institute Journal*, July-August, 75–85.

Shi, C. 1992. Activation of reactivity of natural pozzolan, fly ashes, and slag. Ph.D. thesis. Calgary, Canada: University of Calgary.

Shi, C., Krivenko, P., and Roy, D.M. 2006. *Alkali-Activated Cements and Concretes*. Abingdon, UK: Taylor & Francis.

Shilstone, J. M. 1987. Mix proportions for construction needs. ACI San Antonio Convention, March.

Singh, B. G. 1958. Specific surface of aggregates related to compressive and flexural strength of concrete. *ACI Journal*, 54, 10, 897–907, April.

Solvey, O. R. 1949. *Neue Rationelle Betonerzeugung*. Springer-Verlag.

Standards Australia. 2010. *AS 3972-2010 General Purpose and Blended Cements*. Sydney: Standard Australia.

Stanish, K., and Thomas, M. 2003. The use of bulk diffusion tests to establish time-dependent concrete chloride diffusion coefficients. *Cement and Concrete Research*, 33, 55–62.

Summers, G.R. 2004. A framework for durable concrete. *Concrete Technology Today*, 3, 1, 22–29.

Tang, L. and Sorensen, H.E. 2001. Precision of the Nordic test methods for measuring the chloride diffusion/migration coefficients of concrete. *Materials and Structures*, 34, 479–485.

Tarn, C. T. 1982. Use of superplasticizers to compensate for clay content of fine aggregate. World of Concrete Symposium, Singapore.

Tattersall, G. H. 1991. *Workability and Quality Control of Concrete*. London: E & FN Spon.

Telisak, T., Carrasquillo, R.L., and Fowler, D.W. 1991. Early age strength of concrete: A comparison of several non-destructive test methods. Report No. FHWA/TX-91-1198-1F.

Thomas, M., Folliard, K., Drimalas, T., and Ramlochan, T. 2008. Diagnosing delayed ettringite formation in concrete structures. *Cement and Concrete Research,* 38, 841–847.

Tobin, R.E. 1978. Flow cone sand tests. *ACI Journal*, 75, 1–12.

Toffler, A. 1981. *The Third Wave*. Pan Books.

Torrent, R. 2009. PermeaTORR: Modelling of Air-Flow and derivation of formulae. Materials Advanced Services Ltd. Report, 9 p. Accessed from http://www.m-a-s.com.ar/eng/documentation.php.

Trejo, D., and Pillai, R.G. 2003. Accelerated chloride threshold testing: Part I—ASTM A 615 and A 706 reinforcement. *ACI Materials Journal*, 100, 6, 519–527, November–December.

Trinder, P.W. 2000. Review of the use of Xypex waterproofing and concentrate for CC300 Tsuen Wan Station and Approach Tunnels. Taywood Technical Report No. 1303/00/11129.

Trinder, P.W., Chalmers, C., Peek, A., and Green, W. 1999. Resistance for concrete to harsh environment –Ammonium sulphate. *Concrete in Australia.*

Turner, L., and Collins, F. 2012. Geopolymers: A greener alternative to Portland cement? *Concrete in Australia*, 38, 1, 49–56.

Vallini, D. and Aldred, J.M. 2003. Durability assessment of concrete specimens in the tidal and splash zones in Fremantle port. Coasts & Ports Australasian Conference, Auckland, New Zealand.

Vennesland, O. 1981. Report 3: Corrosion properties. FBC/SINTEF. Trondheim, Norway: Norwegian Institute of Technology, Report STF65 A81033.

VicRoads. 2010. Section 703: General concrete paving.

Vuorinen, J. 1985. Application of diffusion theory to permeability tests on concrete—Parts I and II. *Magazine of Concrete Research*, 37, 132 145–152, 153–161.

Walker, S., and Bartel, F. F. 1947. Discussion of a paper by M. A. Swayze and A. Gruenwald, Concrete Mix Design—A Modification of the Fineness Modulus Methods, Proceedings AI, 43, 2.

Whiting, D. 1988. Permeability of selected concretes. ACI Special Publication, SP 108–11, 195–222.

Xu, H., Provis, J.L., van Deventer, J.S.J., and Krivenko, P.V. 2008. Characterization of slag concretes. *ACI Materials Journal*, 105, 2, 131–139.

Yiannos, P.N. 1962. Molecular reorientation of some fatty acids when in contact with water. *Journal of Colloid Science*, 17, 334–347

Yodmalai, D., Sahamitmongkol, R., and Tangtermsirikul, S. 2009. Chloride resistance of cement paste with crystalline materials. Annual Concrete Conference 6, Thailand.

Index